HVAC TECHNICIAN CERTIFICATION

Exam Guide

SECOND EDITION

By Norm Christopherson

Published by:

DELMAR
CENGAGE Learning

dewalt.com/guides

DELMAR
CENGAGE Learning™

DєWALT: HVAC Technician Certification Exam Guide, Second Edition
Norm Christopherson

Vice President, Technology and
   Trades Professional Business Unit:
   Gregory L. Clayton

Director of Building Trades: Taryn Zlatin McKenzie

Executive Editor: Robert Person

Associate Product Manager: Nobina Preston

Director of Marketing: Beth A. Lutz

Marketing Manager: Marissa Maiella

Production Director: Wendy Troeger

Production Manager: Sherondra Thedford

Manufacturing Buyer: Julio Esperas

Art Director: Benjamin Gleeksman

For product information and technology assistance, contact us at
Cengage Learning Customer & Sales Support, 1-800-354-9706

For permission to use material from this text or product,
submit all requests online at **www.cengage.com/permissions**
Further permissions questions can be emailed to
permissionrequest@cengage.com

ISBN-13: 978-0-9797403-0-5

ISBN-10: 0-9797403-0-4

Delmar
Executive Woods
5 Maxwell Drive
Clifton Park, NY 12065
USA

Cengage Learning is a leading provider of customized learning solutions with office locations around the globe, including Singapore, the United Kingdom, Australia, Mexico, Brazil, and Japan. Locate your local office at **www.cengage.com/global**

Cengage Learning products are represented in Canada by Nelson Education, Ltd.

Visit us at www.InformationDestination.com

For more learning solutions, please visit out corporate website at **www.cengage.com**

Printed in the United States of America
8 9 10 11 12 13   18 17 16 15 14

CONTENTS

DeWALT HVAC TECHNICIAN CERTIFICATION EXAM GUIDE – SECOND EDITION

by
Norm Christopherson

Safety Tech Tip

Carbon monoxide testing should be a regular part of a technician's service. Carbon monoxide is a deadly, poisonous, tasteless, odorless gas resulting from incomplete combustion of a fuel. Called the silent killer, it can also result in severe illness often mistaken for the flu. Some of the symptoms of carbon monoxide poisoning include:

- *Confusion*
- *Dizziness*
- *Headaches*
- *Eye irritation*
- *Nausea*
- *Convulsions*
- *Fatigue*
- *Flu like symptoms*

If you do not measure for carbon monoxide, you do not know it is there. Go to www.bacharach-training.com for more information.

ABOUT THIS BOOK

Congratulations! You have in your hands the best study guide available for HVAC Technician Certification exams. This guide provides a structured exam preparation program to help you master the material that most often appears on all HVAC Certification Exams. It cuts straight to the facts you really need to know to pass any of the series of exams for your area of certification. The book covers all the subjects you will be tested on, including: air conditioning, air distribution, heat pumps, gas heat, and the ever present electrical questions. And it reviews the soft subjects always tested: customer relations, English vocabulary and grammar, safety issues and codes.

For every area of specialty, the book provides an overview along with expert tips and techniques for taking the exams. It discusses exam strategies and making the most of your study time, and offers tips to increase your chances of making the correct choice when you must guess the answer to a question.

In practically no time you will be sailing through confidence-building practice exercises, practical quizzes, and full-length sample exams that will sharpen your skills so you will be ready when test day arrives.

Who Should Use This Book?

HVAC professionals and students alike will benefit by using this book to prepare for any of the HVAC Technician Certification Exams.

This book is for you if:

- You wish to prepare for any of the series of NATE (North American Technician Excellence) exams in the following:

Air Conditioning	**Gas Heat**
Air Distribution	**Heat Pumps**
Core	**Oil Heat**

- You wish to prepare for the HVAC Excellence exams prepared and administered by ESCO (Education Standards Corporation) for the heating and cooling industry, including:

Air Conditioning	**Heating**
Carbon Monoxide	**Heat Pumps**
HVAC Electrical	**Hydronics 1 and 2**

- You wish to prepare for the ICE (Industry Competency Exam), designed primarily for those completing or about to complete a technical program in HVAC at a community college or trade school.

- You are an apprentice preparing for a union journeyman's turn out exam to earn your union journeyman's card.

- You want to take the RSES CM exam or one or more of the Specialist exams that RSES offers.

- You want to increase your chances of gaining employment with a contractor who recognizes the value of technician certification.

- You want a structured guide that explains the exams and takes the guesswork out of preparing for them.

- You want to prepare on your own time and at your own pace.

- You want to prepare for the exams with the advice from an expert instructor who has taken them and knows their content and approach.

- You want a guide that covers all the key areas but does not waste time on topics not in the exams.

This HVAC Technician Certification Exam Guide is designed to improve your test scores no matter which of the many technician certification exams you will be taking!

Troubleshooting Tech Tip

It is necessary to fully understand the normal sequence of operation for the machine being serviced. Then you should be able to determine when in the sequence the machine fails, and what components could stop the operation at that point in the sequence.

Refrigerant Tech Tip

Refrigerant 22 will no longer be used in new systems after 2010 and will no longer be produced as a refrigerant after 2020. Refrigerant 410A is to be the new refrigerant of choice for air conditioning systems.

INTRODUCTION

Reviewing the test strategies and tips in this book as well as taking the sample quizzes and full-length practice exams will give you a big advantage when sitting for any of the HVAC exams. You will learn powerful strategies for answering every type of question which will put you in a better position to obtain a passing grade.

The book's organization suggests a simple study strategy. First, learn the material for your specialty at your own pace and convenience. Next, take one or more of the short 50-question practice exams. Check your answers against the correct answers in the book. When you feel confident that you know the material, take the full-length 100-question practice exam. Your trial performance will help you pace yourself on the actual exam.

The passing score on the exams is 70%. If you achieve this on the practice exams, you should have no problem on the actual exams. You should note that the questions in the practice exams are not the actual exam questions, so do not memorize the questions and answers. Learn the material, pass the exam and become a better technician. Finally, keep in mind that, although scoring high is something to be proud of, passing is the main objective. Once certified, no one knows your scores unless you tell them.

HOW THIS BOOK IS ORGANIZED

Part One is an overview of the testing organizations NATE, RSES, HVAC Excellence, and ICE and their approaches to HVAC exams. It answers the questions *"Why so many different testing organizations?"* and *"Why so many different exams?"*. Other questions such as *"Which exam or exams should you take?"* and *"What do I have to do to maintain my certification?"* are also covered, as well as additional information you should know about technician certification.

Part Two provides test-taking tips, exam strategies, types of exam questions, and how to get the advantage over the exam writers. The author of this book shares his expert knowledge as a professional HVAC instructor and exam writer and shares the secrets of passing exams.

Part Three reveals material you need to know about the so-called "soft studies" exam questions. These are the questions regarding English grammar, vocabulary, customer relations, human resources, honesty and integrity. Yes, these topics appear on the HVAC exams and you need to be prepared for them.

Part Four deals with the math questions that often appear on the HVAC certification exams. Some technicians suffer from math anxiety and tense up just thinking about taking a math exam. This section helps you deal with your worries. Just knowing the type of math questions that most often appear will help relieve most of your concerns.

Part Five supplies the information you need to know to deal successfully with any electrical questions that may appear on an exam. Electricity and electronics are integral to the HVAC service and installation business.

Part Six contains the material encountered most often on the air conditioning exams. Air conditioning centers on the control of comfort of humans, involving the control of space temperature, humidity, air movement, air distribution and filtration. After offering valuable hints, this section provides two 50-question practice exams.

Part Seven covers heat pumps and heat pump exams.

Part Eight deals with gas heat and typical exam questions.

Part Nine addresses the topic of air distribution.

Part Ten prepares you for the topic of controls.

Part Eleven covers the subject of indoor air quality and carbon monoxide.

Part Twelve final practice exams.

Part Thirteen contains answer keys for all the practice exams.

Refrigerant Tech Tip

Liquid and suction lines previously used with R-22 can be used with R-410A provided the lines are clean, in good condition and of the correct size. Oil from the R-22 refrigerant must be purged from the lines.

Charging Tech Tip

Charging a system by clearing the sight glass is never a good practice. Charging a blended refrigerant such as R-410A by sight glass nearly always results in an overcharge of refrigerant. Weigh the refrigerant in, use the manufacturer's charging charts or use the superheat-subcooling method.

OVERVIEW OF THE EXAMS

HVAC Technician Certifications

Four organizations offer technician certifications. They are The Refrigeration Service Engineers Society (RSES), The North American Technician Excellence (NATE), The Industry Competency Exam (ICE) and The HVAC Excellence. Each has a number of specialty exams for those who wish to show their competency in a specific area of technical knowledge and skill. The degree of difficulty varies among them. Currently, these certifications are only voluntary*. They are separate and unrelated to contractor, local, state or national licensing. Technicians who pass any of the exams simply indicate that they have met the tested level of competency.

Technician certification in the HVAC industry had been slow to catch on until many of the major manufacturers began supporting the idea. Several manufacturers of HVAC equipment have begun making it a requirement that a given percentage of their dealer-contractor's technicians become certified. Failure to employ certified technicians could result in the loss of certain dealership privileges. One manufacturer pays more for warranty labor that is performed by a NATE certified technician. Customer education programs are informing customers of the advantage of using contractors who employ certified technicians and contractors are advertising that they employ certified technicians. Therefore, certified technicians should find having one or more certifications will increase their worth to their employers as well as prospective employers.

Each of the certifying organizations has test sites located throughout North America. Exams must be taken in person, and the testing fees vary depending upon the particular certification organization. In the future, exams may be taken over the Internet in the presence of a recognized proctor.

The following is a summary of the major certifications available.

---

*EPA refrigerant certification is a separate certification and is a federal requirement for those who work with refrigerants. This book does not cover EPA refrigerant certification.

- **Refrigeration Service Engineers Society (RSES)**
  RSES is an educational technical society with its headquarters near Chicago and with chapters located throughout North America. Chapters exist in all states and in Canada. A few chapters are located outside the U.S. and Canada.

  RSES gives exams for the following certifications:

  Certificate Member Certification or CM
  Also, RSES gives Specialist Members exams in the following areas:

Air Conditioning	**Controls**
HVAC Electrical	**Domestic Service**
Commercial Refrigeration	**Heat Pumps**
Heating	

  To take any of these exams, one must be an RSES member. Those desiring to take any of the specialist exams must first pass the CM exam. The CM exam covers a broad scope of material and is fairly rigorous. The specialist exams are even more difficult and are considered to be the most comprehensive and challenging of all technician certifications. Only a few thousand technicians have achieved CM status, and fewer still have passed one or more of the specialist exams. Only a handful of individuals have passed all of the RSES specialist exams. Technicians holding specialist status in one or more areas are considered among the elite of the industry.

- **NATE (North American Technician Excellence)**
  NATE states that their certification exams test on what 80% of all technicians need to know 80% of the time. NATE requires that those certified with them take continuing technical education seminars or short courses in order to maintain certification.

  NATE offers exams and certifications in the following areas:

Air Conditioning (Installation)	**Gas Heating**
Air Conditioning (Service)	**Heat Pumps**
Air Distribution	**Oil Heating**

- **ICE (Industry Competency Exam)**
  Administered under ARI, the Air Conditioning Refrigeration Institute, this certification is entry level (very basic) and is often utilized by trade schools, vocational schools and community colleges as a requirement to graduate from HVAC educational programs.

Refrigerant Tech Tip

HFC refrigerants such as R-410A and R-134A use only POE oils. POE oils provide better oil return, have improved heat transfer characteristics and lubrication and are wax free. However, POE oils are highly hygroscopic, must be kept in sealed metal containers and transferred with a pump. Never leave a POE system open to the atmosphere any longer than necessary. Replace the filter-drier and pull a 500 micron vacuum. Do not take shortcuts with systems containing POE oil.

Tool Tech Tip

Always use a ratchet wrench on the stem of a service valve. Adjustable wrenches will round off the flats on the end of the stem, making it impossible to operate the stem later. Clamp-on type wrenches will score the shaft of the stem and tear up the valve packing, causing a packing leak. Some service valves are welded to the compressor or receiver making it impossible to replace the valve without replacing the entire compressor or receiver.

- **HVAC Excellence**
  (A subsidiary of ESCO: Educational Standards Corporation)

  Air Conditioning Certification Core Exam

  Electric Certification Core Exam

  Carbon Monoxide Certification

  Commercial AC Certification

  Commercial Refrigeration Certification

  Electric Heating Certification

  Gas Heat Certification

  Geo Thermal Certification

  Heat Pump Certification

  Hydronics 1 (Steam) Certification

  Hydronics 2 (Hot Water) Certification

  Oil Heat Certification

  Refrigerant 410A Certification

WHY BECOME CERTIFIED?

Thousands of technicians have become certified in one or more areas. Many hold certifications from more than one testing agency. The number of certified technicians has grown rapidly since the year 2000, and the number of individuals taking the exams is setting new records. Why do they bother, and why should you?

- Certification is becoming the benchmark for contractors who need a consistent method of judging a prospective employee's knowledge and ability. Certification proves to some degree that you are competent in the area in which you are certified.

- Equipment manufacturers are encouraging their distributors and dealers to employ certified technicians or get most of their current technicians certified. The manufacturers are taking a three-pronged approach to technician certification. First, they are endorsing one or more of the certifying exam agencies. Second, the manufacturers are providing their distributors and dealers with special incentives for certifying a certain percentage of their technicians. In some cases, they are even assisting with the costs of certification. Third, some manufacturers may penalize those who fail to make progress at certifying or employing certified technicians.

- Some state and local governments are beginning to use technician certifications as requirements for obtaining state and local licenses. At least one school district is requiring its current air conditioning and heating technicians to obtain certification and will use certification as a component of the hiring procedure for future openings.

- Some branches of the United States Armed forces are accepting NATE technician certification and in some cases are requiring it for advancement.

- A number of local union apprenticeship training programs have already incorporated one or more of the technician certification exams into their journeyman's programs. At the national level, the unions are seriously considering doing the same, which will require the participation of every local union.

- It is very likely that insurance companies providing coverage to HVAC contractors will soon recognize that trained and certified technicians lower the risk of incidents where insurance claims are filed, and begin offering discounted insurance rates to contractors with certified technicians. This in turn will further encourage the manufacturers and contractors to employ certified technicians.

- Certifications are not only of value to your employer, they are of personal value. Remember, it is not so much the certifications themselves that are so important, but the knowledge, skill and new information that you gain in the process of obtaining any particular certification.

- Manufacturers, certifying agencies and contractors are targeting consumers in advertising campaigns that focus on the value of using contractors who employ certified technicians. The days of the "backyard" untrained HVAC guy are coming to an end, much as they have for the automobile repair business. HVAC systems are complex and require highly skilled technicians with the proper diagnostic and service gear to maintain them properly. Certification is one more way for the customer to differentiate between contractors.

The entire HVAC Technician Certification movement is changing so rapidly that by the time you hold this book in your hands, many more technicians will have taken one or more of the exams, more manufacturers will have endorsed one or more of the certifying agencies, and more HVAC employers will be using the certifications as an employment tool.

Safety Tech Tip

Always remove and replace pulley belts by loosening the motor and sliding it forward. Always use the palm of your hand on the belt, keeping your fingers off the belt and pulleys, as even a strong wind can cause a pulley to rotate and cut your fingers.

Charging Tech Tip

Always make sure a system is properly charged with the correct amount of refrigerant. An overcharged system can cause liquid flood-back, flooded starts, loss of lubrication, loss of capacity, high operating costs, high head pressure and premature system failure.

STATE CONTRACTOR LICENSING EXAMS

Do not confuse technician certification with state contractor licensing exams. Many states require these in order to offer contracting services legally in that state. State contractor licensing exams vary widely among the states. However, they usually have many things in common. State licensing exams for HVAC often include questions on heating and cooling theory, system construction, operation, installation, safety and calculations. State contractor exams also consist of business, accounting, estimating, tax code, legal issues and understanding the mechanical and electrical codes.

This book is a valuable study guide to the heating and cooling theory, system construction, operation, installation, safety and calculation portions of the state licensing exams.

THOUGHTS ON MEMORIZATION, EQUATIONS AND CALCULATORS

Let's face it. No matter what your instructors have told you about not memorizing material, we learn by repetition and review. Certain material simply must be memorized. You learned your multiplication tables through pure memorization. You know there are 12 inches in a foot and three feet in a yard by memorization. You should know that there are 746 watts per motor horsepower and a gallon of water weighs 8.33 lbs.

Some information is memorized and recalled when needed. Other information can be reasoned or figured out by experience, logic and common sense. However, your ability to use logic and common sense often requires some background knowledge that was memorized.

In this study guide I have included a section containing critical items that simply require memorization. The NATE exams have all the HVAC math equations related to the exam printed inside the front and/or back cover of each exam. There is no need to memorize equations. However, you need to know how the equations work, how to apply them to solving problems and be able to recognize which equation is required to solve a particular problem. On the job you are allowed to look up information, refer to an equation and use a calculator. So should you be able to do so on the examinations. Both HVAC Excellence and NATE allow calculators. No notes or reference materials are allowed. Any math equation necessary will be furnished for you. Just make sure you are familiar with the equations and equally familiar with your calculator. My favorite calculator is the Texas Instruments TI-30. It is not expensive, is easy to use and has all the functions necessary to meet the needs of a technician.

STATE LICENSING REQUIREMENTS FOR HVAC CONTRACTORS

State	Licensing Board Phone Number	Licensing Board Website	State Licensing Exam/ Testing Company	Pre-licensing or Pre-approval	Continuing Education
Alabama	(334) 242-5550	www.hvacboard.state.al.us	Yes/PSI	Yes/Pre-approval	Yes
Alaska	(907) 465-3035	www.dced.state.ak.us/occ/pcon.htm	Yes/Thomson Prometric	No	Yes
Arizona	(602) 542-1525	www.rc.state.az.us/	Yes/Thomson Prometric	No	No
Arkansas	(501) 661-2642	www.healthyarkansas.com	Yes/International Code Council	Yes/Pre-approval	No
California	(800) 321-2752	www.cslb.ca.gov	Yes/California State Licensing Board	Yes/Pre-approval	No
Colorado	Not Regulated by State/contact city/county		No		
Connecticut	(860) 713-6135	www.state.ct.us/dcp	Yes/PSI	Yes/Pre-approval	No
Delaware	(302) 744-4504	www.dpr.delaware.gov	Yes/TBA They're in the process of setting up the exam	Yes/Pre-approval	No
Florida	(850) 487-1395	www.state.fl.us	Yes/Professional Testing Inc.	No	Yes
Georgia	(478) 207-1416	www.sos.state.ga.us/plb/construct/	Yes/AMP	Yes/Pre-approval	Yes
Hawaii	(805) 586-2689	www.hawaii.gov/dcca/pvl	Yes/Thomson Prometric	Yes/Pre-approval	No
Idaho	(208) 332-7158	http//dbs.idao.gov/hvac.licence.html	Yes/International Code Council	Yes/Pre-approval	No
Illinois	Not Regulated by State/contact city/county		No		
Indiana	Not Regulated by State/contact city/county		No		
Iowa	(515) 251-7995	www.iowaworkforce.org/	No/Register with State; City/County may require testing		
Kansas	Not Regulated by State/contact city/county		No		
Kentucky	(502) 573-0395	http://www.ohbc.ky.gov	Yes/International Code Council	Yes/Pre-approval	No
Louisiana	(504) 825-2382	www.lslbc.state.la.us	Yes/State Licensing Board	Yes/Pre-approval	No
Maine	Not Regulated by State/contact city/county		No		
Maryland	(410) 230-6159	www.dllr.state.md.us	Yes/PSI	Yes	No
Massachusetts	(617) 727-9931	www.state.ma.us	No	Yes/Pre-approval	No
Michigan	(517) 241-9325	www.michigan.gov/dleg	Yes/State Mechanical Board	Yes/Pre-approval	No
Minnesota	Not Regulated by State/contact city/county		No		
Mississippi	(601) 354-6161	www.msboc.state.ms.us	Yes/PSI	Yes/Pre-approval	No

STATE LICENSING REQUIREMENTS FOR HVAC CONTRACTORS (CONTINUED)

State	Contact	Website	Exam	Pre-approval	
Missouri	Not Regulated by State/contact city/county		No		
Montana	Not Regulated by State/contact city/county		No		
Nebraska	Not Regulated by State/contact city/county		No		No
Nevada	(702) 486-1100	www.nscb.state.nv.us	Yes/PSI	Yes/Pre-approval	No
New Hampshire	Not Regulated by State/contact city/county		No		
New Jersey	Not Regulated by State/contact city/county		No		
New Mexico	(505) 452-8311	http://www.rld.state.nm.us/CID/index.htm	Yes/PSI	Yes/Pre-approval	No
New York	Not Regulated by State/contact city/county		No		
North Carolina	(919) 875-3612	www.nclicensing.org	Yes/PSI	Yes/Pre-approval	No
North Dakota	(701) 328-9979	http://www.legis.nd.gov/information/rules/admincode.html	No/Register with State		
Ohio	(614) 644-3493	www.com.state.oh.us/dic/dicocilb.htm	Yes/International Code Council	Yes/Pre-approval	Yes
Oklahoma	(405) 271-5217	www.cib.state.ok.us/	Yes/PSI	Yes/Pre-approval	Yes
Oregon	(503) 378-4133	www.oregonbcd.org	Yes/State Plumbing Board	Yes/Pre-approval	No
Pennsylvania	Not Regulated by State/contact city/county		No/city/county may require testing		
Rhode Island	(401) 462-8000	www.dlt.state.ri.us	Yes/State Electrical Board	Yes/Pre-approval	No
South Carolina	(803) 896-4686	www.llr.state.sc.us	Yes/PSI	No	No
South Dakota	Not Regulated by State/contact city/county		No		
Tennessee	(615) 741-8307	www.state.tn.us/commerce/boards/contractors/	Yes/PSI	No	No
Texas	(512) 456-2145	www.tsbpe.state.tx.us	Yes/PSI	Yes/Pre-approval	Yes
Utah	(801) 530-6628	www.dopl.utah.gov	Yes/Thomson Prometric	Yes/Pre-approval	No
Virginia	(804) 367-8511	www.state.va.us/dpor	Yes/PSI	Yes/Pre-licensing	No
Washington	(360) 902-5207	www.lni.wa.gov	Yes/LaserGrade	Yes/Pre-approval	Yes
West Virginia	(304) 558-7890	www.labor.state.wv.us	Yes/Thomson Prometric	No	No
Wisconsin	(608) 261-8500	www.commerce.wi.gov	No Exam/Apply to board for Credentials	No	Yes
Wyoming	Not Regulated by State/contact city/county		No		

PART ONE
Certification Organizations and Their Examinations

What You Will Learn

- Why certification is important to you

- The certifying organizations

- The certification exams offered

- Why there are so many different certifying organizations

- Why there are so many different exams

- How to choose which organization to test with

- How to choose which certifications to obtain

- Maintaining your certification

- What to bring and not bring to the examination

Certification Tech Tip

Become a certified technician in each of the areas in which you work. Encourage others to do so as well. Take recognized continuing education classes and seminars to stay current with changes in the industry. New technology and products are developing regularly. Share your new knowledge and skills with others and learn from others as well.

WHY CERTIFICATION IS IMPORTANT TO YOU

By purchasing this book, you have already indicated that certification is important to you. Perhaps your employer has decided that a certain percentage of employees will become certified, and you are one of them. Maybe one or more of the manufacturers your company represents have offered incentives for technician certification. Perhaps you work for a contractor whose new liability insurance policy allows for a substantial discount on the premium if he employs certified technicians. Or customers looking for a heating and cooling contractor have been asking if your company employs certified technicians.

You may live in a state, county or city that has, or is considering, requirements for technician certification in order to qualify for a license to install and service mechanical and electrical heating and cooling systems.

You may be a member of a branch of the armed forces or employed by the federal government as a heating or cooling technician. If you are on a point system for career advancement, it is probable that technician certification will count towards a move up the career ladder.

Many heating and cooling instructional programs at community colleges, vocational schools and trade schools are using the ICE exams as the benchmark for meeting graduation requirements. Union apprenticeship training programs are using the nationally accepted certifications to determine if and when fourth and fifth year apprentices are ready to become full journeymen.

Contractors are realizing that hiring new employees who hold one or more certifications makes it easier to determine how much the job applicant already knows. The nationally recognized exams give the contractor a well-established set of criteria upon which to make his evaluation.

National technician certification is here and is rapidly growing. Thousands of technicians have already been certified in at least two areas of specialty, and the rate at which more technicians are sitting for the exams is at a record high. Technician certification is not a matter of *"Should I become certified?"*; it is a matter of *"Which certifications do I need?"*

THE CERTIFYING ORGANIZATIONS AND THE EXAMS OFFERED

There are three major certifying organizations offering exams. They are the Refrigeration Service Engineer's Society (RSES), North American Technician Excellence (NATE), and HVAC Excellence. Let's take a closer look at each of them.

THE REFRIGERATION SERVICE ENGINEER'S SOCIETY (RSES)

This organization is the "granddaddy" of all the certifying organizations. Established as a not-for-profit educational organization for the professional development of those employed in the heating, ventilating, air conditioning and refrigeration industry, RSES has done a very commendable job with its mission. Now an international organization, it is the service and installation technician's equivalent to the American Society of Heating, Refrigerating and Air Conditioning Engineers (ASHRAE), which is the design and application engineer's organization. RSES has its headquarters near Chicago, and maintains hundreds of chapters around the world with the majority located in North America. Most states and Canadian provinces have local chapters where monthly educational meetings take place. In addition, members attend regularly scheduled training courses held at the chapter level, with the curriculum supplied by the international headquarters. Technical seminars are held regularly across North America, with speakers from well-recognized manufacturers and organizations providing expert instruction. Members also receive a monthly magazine which keeps them updated on industry trends and educational opportunities.

- **RSES – Certificate Member Certification**

  A central part of RSES' mission is the series of technician certifications it offers. Members may take a fairly difficult exam covering the broad scope of technical and practical knowledge required of those in the heating and cooling industry. Those who pass the exam are classified as Certificate Members and are entitled to place the letters CM after their name. Certificate Member status is truly a credit to the level of knowledge the individual has attained. The information covered in this exam preparation guide is very important to anyone about to take on the challenge of passing the RSES CM exam.

A Point To Ponder

"There is no substitute for experience and experience is no substitute for study. Wise men do not condemn studies, they utilize study to make the most of their experience."

Safety Tech Tip

Special non-conducting shoes and boots are available for special and daily wear at work. Shop for footwear that is rated for use around electrical systems. Your feet are the most direct path to ground and we often work on wet roofs and around moisture.

- **RSES – Specialist Exams**

Those who have successfully passed the CM exam may choose to take one or more of RSES' Specialist exams. These series of exams test the individual in an area of specialty to a greater degree of depth and detail than the more general CM exam. Currently, there are specialist exams in the areas of:

Air Conditioning	**Controls**
HVAC Electrical	**Domestic Service**
Commercial Refrigeration	**Heat Pumps**
Heating	

These exams are perhaps the most difficult and comprehensive in the industry, and RSES constantly updates and changes the exams to stay abreast of the latest industry developments. Earning one of these levels of specialty is quite an achievement. Very few individuals have taken and passed more than one of the specialty exams. There is no doubt that someone who has reached the level of obtaining two or more specialist levels is one of only a handful of people in the business.

Although this study guide is a very important tool in preparing for any of these exams, it is recommended that you pursue the RSES courses offered at the local level. In addition, at least five to seven years of industry experience are usually necessary.

The exams include questions requiring written paragraphs, filling in the blanks, drawing sketches and completing wiring diagrams as well as solving more complex technical math problems. It is recommended that you successfully pass several of the other technician certifications in preparation for these exams.

NORTH AMERICAN TECHNICIAN EXCELLENCE (NATE)

NATE is a private organization that is supported by most of the major manufacturers of HVAC equipment as well as by RSES. Many government agencies, unions, contractors and individuals in the HVAC industry endorse NATE and it is well accepted as the primary examining agency of the industry. NATE offers a series of certifications which are all multiple-choice exams. Not quite as difficult as the exams offered by RSES, NATE states that their exams contain questions that test for what 80% of all technicians need to know 80% of the time. The exams are more practical and contain technical material to determine that a technician is knowledgeable in the proper installation and service of heating and cooling systems. NATE requires that a technician pass a core exam of 50 multiple-choice questions in addition to an area of specialty of the technician's choice consisting of an additional 100 questions. The core exam is only taken once, after which the technician may take any one or more of the specialty exams. NATE states that the pass rate for those who do not prepare for the exams through a study guide such as this one is not nearly as high as for those who prepare. This study guide with practice exams is an excellent source for preparing for the NATE series of exams.

NATE offers different specialty exams for installers and service technicians. Those who pass the service technician exams automatically receive credit for passing the installers exams. Therefore, this guide concentrates on preparing you to pass the service technician exams rather than having separate practice exams for installers and service technicians.

There are different versions of each NATE specialty exam, so a group of people taking an exam on the same specialty take different versions of the same exam. This helps prevent cheating and allows those who fail an exam to retest with a different version.

All the NATE exams allow the technician to use a scientific calculator and scratch paper. Any math equations required to answer test questions are printed on the inside cover of the test booklet. The technician does not have to memorize equations but must be capable of determining which equation is necessary to solve the question and properly utilize the equation.

- **The NATE Core Exam**
  The core exam tests the technician for his or her knowledge in the areas of reading skills, basic math, vocabulary, spelling, simple grammar, customer relations, human resources, and honesty and integrity. This study guide covers the key items tested in the core exam as well as the key material tested in the NATE specialty exams. The core exam is taken at the same time as one of the NATE specialty exams. You first take the core exam and then the specialty exam of your choosing. Some technicians choose to take the core and two specialty exams at one sitting.

- **The NATE Air Conditioning Specialty Exam**
  Topics on this exam include basic theory of heat transfer, basic electrical concepts, applied electricity, electrical and refrigerant troubleshooting, tools and materials, installation and servicing procedures, generally accepted codes and safety. The majority of the questions are aimed at residential and light commercial applications and are of the multiple-choice type.

Learning Tech Tip

Always learn new skills under the supervision of an experienced and qualified service technician who has performed the operation successfully in the past. Read and follow the manufacturer's installation and service instructions even if you have performed the service before. New technology requires new knowledge, skills and methods of service.

Building Pressure
Tech Tip

Infiltration is air that leaks into a building due to lower pressure in the building than that outside the building. Exfiltration is air that leaks out of the building due to higher pressure in the building than that outside the building. A slightly positive interior pressure is desired.

- **The NATE Heat Pump Specialty Exam**

  This exam concentrates on material that a technician should know in order to understand and properly install and service heat pump systems. The majority of the exam deals with residential and light commercial air-to-air heat pumps. This study guide provides a heat pump technician with the review necessary to make a good showing on the heat pump specialty. The technician who passes the NATE Heat Pump Specialty automatically receives credit for having passed the Air Conditioning Specialty and the installer specialty as well. That alone makes this particular specialty a good one to achieve.

- **The NATE Gas Heat Specialty Exam**

  This specialty covers what technicians need to know when installing and servicing gas heating systems. The majority of the material on the exam deals with residential and light commercial equipment. The exam includes the topics of proper installation, sequences of operation, heating system components, special considerations on higher efficiency systems, the gas train, the electrical system, code considerations and troubleshooting.

- **The NATE Oil Heating Specialty Exam**

  The oil heating specialty exam covers a wide variety of topics closely related to what technicians need to know in order to successfully install and service oil heating systems. As with the other NATE exams, this exam also deals primarily with residential and light commercial oil heating systems.

- **The NATE Air Distribution Specialty Exam**

  This specialty exam certifies that the technician has the knowledge to properly choose, install and service air duct systems and components utilized in residential and light commercial heating and cooling applications. Topics in this specialty include the proper selection, care and use of sheet metal tools, methods of constructing and joining ductwork, types of air distribution systems, related codes, sheet metal workplace safety, selection and placement of registers and grilles, airflow measurement, airflow test instruments and work processes.

HVAC EXCELLENCE

HVAC Excellence is the third, but not the least important, of the three major technician certification organizations. HVAC Excellence is a subsidiary of Educational Standards Corporation (ESCO), an organization that develops and provides tests for a variety of industries. They employ professionals who write exams with the help of industry consultants. Then they administer the exams, correct and score them, and maintain a list of those who are successful. HVAC Excellence offers exams and certifications in air conditioning, heating and heat pumps, and specialties not offered by other certifying agencies including a carbon monoxide certification with others yet to be made public. This study guide will help you prepare to pass the majority of the specialty certifications offered by this organization. HVAC Excellence uses electrical certification as their core exam. You must take and pass their electrical exam in order to gain any of the other certifications. Each exam is made up of 100 questions. The HVAC Excellence exams are composed of multiple-choice questions, as are the practice exams in this study guide. A passing score for all the exams is 70%.

INDUSTRY COMPETENCY EXAMS (ICE)

ICE is the easiest of the exams offered. ICE is designed and administered by ARI, the Air Conditioning Refrigeration Institute. The ICE exams are intended to test students in community colleges, vocational schools and trade schools who are about to graduate from those programs. ICE serves several purposes. ICE allows schools to determine that they are providing the instruction required to gain entry into the heating and cooling industry. ICE indicates to contractors that the graduate has met a minimum level of knowledge and understanding necessary to start a successful career. Finally, passing the ICE exams indicates to the student that he or she has met the industry's accepted entry-level standards. Currently, there are three ICE exams: Residential air conditioning, light commercial air conditioning, and commercial refrigeration. This guide provides instruction for students facing this first level of certification, and it is a vital reference for preparing for the remaining levels of certification necessary to a rewarding career.

Instrument Tech Tip

Precision is a measure of how small a value an instrument can measure. Accuracy is a measure of how close an instrument will read to the true value. Resolution is the least change in value to which the instrument will respond. Repeatability means obtaining the same reading time after time under the same conditions.

A Point To Ponder

If you do not know how it works when it works right, how will you know when it is fixed? To be a better technician, learn how the equipment works!

WHY ARE THERE SO MANY DIFFERENT CERTIFYING ORGANIZATIONS AND EXAMS?

These are questions commonly asked in the HVAC industry community, and there are no clear answers. We may eventually see several of these technician certification organizations combine into one larger organization. That has happened to other organizations in the HVAC industry over the years. What will happen to your certifications if several organizations merge? We will most likely see grandfathering of certifications at the time of a merger. I hold a certification from RSES that automatically gave me several certifications with NATE without having to take the NATE exams. However, I took the NATE exams anyway.

The question of why there are so many different exams is more easily answered. Simply, the HVAC industry is a broad and complex one with a number of specialties. Some technicians specialize in controls or air and water balancing. Others specialize in hydronic heating, while some specialize in still other areas. There are contractors who only perform particular aspects of heating and cooling and they employ technicians who specialize in that particular aspect of the business. For example, there are contractors who only sell, install and service computer room air conditioning. Others concentrate on ice machines. Still others specialize in controls. A growing area of specialization is indoor air quality. Look for more technician certifications in these and other areas of our industry. It is an unfortunate fact that there is an urgent need for more highly trained technicians than are currently available to fill the openings. Obviously, there are opportunities for anyone interested in entering the industry or for those already in the business to increase their value through continuing education and certifications.

HOW TO CHOOSE AN ORGANIZATION TO TEST WITH AND HOW TO CHOOSE WHICH SPECIALTIES TO OBTAIN

Having read this far, you likely have a pretty good idea of which specialties best suit your particular need. Obvious questions to answer are what aspect of the business you work in and what aspect of the business you want to work in. If you are employed as an installer and you want to move into a service position within your company, it may be best to choose a certifying organization that is recognized by the company you work for or endorsed by the brand of equipment your company is associated with. Discuss your options with those you work with, your employer or an employer you wish to work with. Your choice of testing organizations may be limited by the particular certification you are seeking. Technical school students may be required to take one or more of the ICE exams for graduation. No matter which testing organization you choose or which certification you seek, utilizing this study guide will definitely increase your exam score.

MAINTAINING YOUR CERTIFICATION

RSES

RSES CM and Specialist certifications are good for life, never requiring the holder to retest or take any continuing education classes, seminars or complete additional work as long as the holder maintains his or her membership in RSES.

NATE

NATE certification is valid for five years. At the end of five years, it must be renewed or it is lost. There are several ways to renew a NATE certification:

1. You may do nothing during the five years and then retake the certification exam.

2. You may accumulate 30 hours of NATE-recognized* continuing education and then take a shorter 50-question exam in your particular certification.

3. You may accumulate 60 hours of NATE-recognized* continuing education and not have to take the exam.

HVAC Excellence

HVAC Excellence certifications are good for life and never require the holder to retest or take any continuing education classes or seminars.

ICE

ICE certification is not renewable as it is expected that the holder will move beyond entry-level certification after gaining a few years of industry experience.

---

*NATE-recognized training is training that NATE has specifically approved as acceptable for renewing your certification. Only NATE-approved training is acceptable.

Customer Tech Tip

The customer does not know how technically sharp you are. The customer will determine your ability by how neatly you work, how well you communicate, how friendly you are, how you treat him or her and by your overall presentation. Keep your service vehicle clean, wear a clean uniform, listen to the customer and look him or her directly in the eye. Every service call is a sales call. The technician has a great deal of control over repeat business.

Customer Tech Tip

Never argue with a customer even when the customer is unhappy or upset. Remain calm and professional. Listen to the complaints or problems and find alternatives that are acceptable to your company and the customer.

WHAT TO BRING AND NOT BRING TO THE EXAMINATION

On the day of the scheduled examination there are several items you are required to bring. All of the examination organizations require photo identification such as a valid driver's license. You must also be prepared to pay the examination fees if you have not made arrangements beforehand. Bring several No. 2 lead pencils as well. A non-programmable scientific calculator is allowed and will be necessary. With the possible exception of the RSES CM and Specialist exams, any math equations necessary are provided. Refrigerant saturation or temperature-pressure charts are also provided.

You are not allowed to bring any reference materials, books, notes or even your own blank scratch paper into the examination room. Anything you are allowed to use on the exam will be provided by the examiner. You will be monitored throughout the examination to ensure the honesty and integrity of the testing process.

For details on what will be allowed, contact the organization prior to your scheduled test date.

PART TWO
Exam Strategies

What You Will Learn

- General test-taking strategies

- Preparing for an exam

- How to study effectively

- What to study

- Common strategies for success with multiple-choice questions

- Tactics used by test writers

- Evaluating multiple-choice questions

Study Tech Tip

Do not object to reviewing material you think you already know. Often you will learn something new or correct a misunderstanding.

GENERAL TEST-TAKING STRATEGIES

Do not make the exam harder than it is by failing to devote an adequate amount of time to preparation. Set aside regular time for study and stay with your schedule. Study in a quiet and well-lighted place with room for your study materials. Keep the top of your desk clear of all unrelated material and concentrate on the task at hand. Do not wait until the last minute and then "cram" for the exams. Plan your time and study regularly over a period of months or weeks before the exam date.

Take all of the practice exams in this guide. Do not skip any, even if they do not directly relate to the particular specialty you will test on. Each practice test provides additional experience with taking multiple-choice exams in general. The practice will help you raise your score on the actual exam.

Do not look at the correct answers before answering the practice questions on your own. This can cause you to think you actually know the correct answers when you do not.

After taking a practice exam, correct and score the exam using the answer key in the back of this guide. After correcting and scoring a practice exam, review every missed question. Read any comments that may be provided with the answer for that practice exam. The comments were composed as an additional aid to help you understand why you missed the question so you do not miss a similar one.

Also review those questions you answered correctly. A great deal can be learned from the process of taking and reviewing an exam. Make the most of the practice exams as a method of study. When I taught college classes in HVAC, I gave quizzes and exams on a regular basis and used each one of them as another method of reviewing the course material. You can do the same using the practice exams.

Give special attention to your weak areas. It may be tempting to spend a great deal of study time perfecting what you are most comfortable with or what you know best. Reviewing your favorite technical topic is useful, especially if you are about to take a specialty exam in your best subject. However, the most valuable study is accomplished in your weakest areas where the greatest gains will occur when you take the actual exams.

Reading is not the same as studying. As you study, evaluate what you are trying to learn. Consider new ways of approaching the topic or solving a problem. Relate new material to what you already know to be correct. If you have a technical library of your own, research the topic under study by reading about it from several sources. Each book may approach the topic from a slightly different angle, or may cover the topic in greater detail. Make notes on new things you learn about the subject and review them later. Discuss items of particular interest with your colleagues at work or school. Often someone may come up with an idea that sparks another to develop it still further.

Do not object to reviewing material that you think you already know well. Do not fall into the trap of thinking "I already know this stuff." Every time I study material I am sure I know, I learn something new, correct a misunderstanding or make a new association. Remember, everything you learn is somehow connected to something you can learn later and add to your storehouse of knowledge. Even after years of teaching HVAC as a profession, I can learn something new about a topic I have taught dozens and dozens of times. Occasionally, a student will ask a question that causes me to rethink a topic from another angle and we both learn something new.

Keep a pencil in your hand while you are studying. Make sketches, redraw basic diagrams, take notes, make lists and underline important passages. Do not be afraid to write useful notes in the margins. After all, the book is yours and for your personal education. I have a number of favorite books that are full of notes I have made. Those books are often the most useful and valuable in my technical library. And, just keeping that pencil in your hand helps you feel serious about studying.

Do not get up from your work until you have accomplished the task at hand and have spent serious time in study and gained from the time invested. Once you have established good study habits, further study comes easier. Review your work constantly. Never pass up a review because you feel you have already mastered a topic.

Study Tech Tip

Do not look at the answers to the practice questions until you have completed the entire sample exam, or you may be led to believe you knew the correct answer in the first place. A test is a great study aid if you allow it to correct your misunderstandings.

Study Tech Tip

Do not wait until the last minute and "cram" for the exams. Study regularly and review items you got wrong at your last study session.

COMMON STRATEGIES FOR SUCCESS WITH MULTIPLE-CHOICE QUESTIONS

The ICE, NATE and HVAC Excellence series of technician certification exams all have one major factor in common. Each consists exclusively of multiple-choice questions. The key to performing well on multiple-choice exams is excellent reading skills. Since reading is central to success with this type of question, your preparation should include a lot of attention to reading multiple-choice questions. Review the following list of common strategies for success with multiple-choice questions.

- Read every word of the instructions. Misreading causes the most damage to your score next to misreading the question itself.

- Carefully read every word of each question. Many multiple-choice questions are missed due to failure to carefully and accurately read the question. Do not hurry through the questions even if you think you know the correct answer.

- Mark your answers on your answer sheet carefully. Completely darken in the answer space. Poorly marked answers may cause the machine-scored answer sheet to score a correct answer as incorrect. It is your duty to mark your answer sheet correctly.

- Pace yourself carefully in order to allow plenty of time to complete the exam. The full-length 100-question practice exam will allow you to time yourself while taking it. Record your starting and ending time for the test. You will probably find you can easily pace yourself to finish in plenty of time to recheck your work.

- Leave difficult questions for later. Do not spend a great deal of time trying to solve a problem you find particularly difficult. Your time is better spent answering easier questions correctly. Each question carries the same point value, so spending a great deal of time on difficult questions could cost you many points if you run out of time and leave other questions unanswered.

- Answer every question even if you have to guess. If you are running short on time do not leave any questions unanswered. Unanswered questions count the same as incorrect answers on these exams, so guessing can only raise your score. However, if you follow the advice of this guide and properly prepare for the exams using the practice exams, you should be able to complete the actual exam with plenty of time to review your work.

- Spend any extra time checking your answers.

- Do not mark more than one answer for each question. Doing so will automatically score that question as incorrect.

- Try not to skip any questions. If you wish to come back to a particular question later, place a mark next to the question so you can locate it. When you skip a question, you run the risk of skipping a line, so all the questions answered after that are recorded on the wrong line. That can cause you to waste a great deal of time erasing and redoing those answers. So be very careful when recording your answers, especially when you intend to skip a question and come back to it later.

TACTICS USED BY TEST WRITERS

Well-written exams have a strategy to them. Experienced test writers know how to construct questions that fool those who are not careful about how they approach the exam.

The scope, contents and style of an examination depend upon the writer. All examinations are of limited scope. An air distribution examination is limited to questions on air distribution. There are only so many questions that can be composed on the same subject without duplication. Variations on a theme may be diverse but the theme remains the same. Where problem solving and math is required, the given conditions may vary greatly, but the basic steps for finding the solution remain unchanged. Learning to solve problems involving the fan laws or Ohm's law are all pretty much the same no matter how the question is asked. Remember there are universal truths, such as: there are 12,000 btuh per ton, 3.42 btu per watt and $E = IR$. Once learned and understood, these truths can only be examined a few ways no matter how creative the test writer. The truths themselves cannot be changed and there are only so many ways to ask the same question. The key is to read the question carefully and make sure you understand what is asked.

The ICE, NATE, HVAC Excellence and RSES exams were written by professional exam writers with the aid of members of committees. The committee members are all experts from industry, usually a combination of engineers from manufacturers, industry association representatives, college teachers, technical writers and mechanical or electrical code inspectors. Each tends to slant exam questions to their own discipline. A college teacher will weight the exam with many theoretical questions, a code official with many code questions and a contractor representative with more practical questions. The best exams are those composed of a variety of questions covering the broad spectrum of the area of specialty. The committee serving on the NATE question construction and review board consists of an excellent mix of representatives from all aspects of the HVAC industry.

Study Tech Tip

Give special attention to areas that are your weakness. Do not be tempted to spend a great deal of study time perfecting an area you are already strong in.

Building Pressure Tech Tip

Infiltration is air that leaks into a building due to lower pressure in the building than that outside the building. Exfiltration is air that leaks out of the building due to higher pressure in the building than that outside the building. A slightly positive interior pressure is desired.

EVALUATING MULTIPLE-CHOICE QUESTIONS

The ICE, NATE and HVAC Excellence exams are all multiple-choice. The question is presented and you are to choose the correct answer from a set of four or five that are provided. Very often all of the answers seem reasonably correct. However, only one is the correct answer. When a skilled examiner has written the multiple-choice question, determining the correct answer can be a real challenge.

Learn these strategies for evaluating multiple-choice questions and your exam scores will soar.

Look for the qualifying word in the question.

Qualifying words are key words that define exactly what the question is really asking. Examples of qualifying words are:

- Always
- Never
- Usually
- Least
- Most
- Most often
- Least often
- Largest
- Least likely
- Most likely
- Probably
- Possibly
- Worst
- Smallest
- Minimum
- Chief
- Most advisable
- Greatest
- Best
- Not
- Maximum

These qualifiers offer a clue to the correct answer.

Carefully observe the way the question is worded.

One question may ask, "Which of the following is not...", and then list four items. Carefully note the key word is "not." You are looking for the answer from the list that does not fit what was asked. The very next question may reverse the process and ask for something that fits the question. Careful reading of the question is essential. Here is an example:

1. Which of the following tools or instruments are not used during a system evacuation?
   a. A vacuum pump
   b. A micron gauge
   c. Refrigerant hoses
   d. An anemometer

2. Which of the following tools or instruments are used when evacuating a system prior to charging?
   a. An inclined manometer
   b. An electronic micron gauge
   c. A megohm meter
   d. A calculator

Do not assume the meaning of a word.

Some questions are missed because the examinee did not know the meaning of a word in the question or the meaning of a word in the answers provided. As you study, have a dictionary nearby and look up the meaning of any word you do not fully understand. Also, have an HVAC dictionary or glossary handy for industry specific terms such as The DEWALT HVAC Professional Reference Guide.

If you are unsure of the answer, apply the principle of elimination.

First, eliminate the answer or answers that are obviously incorrect. This narrows your choices to perhaps two or three. If you can reduce the choice to two, your chances of selecting the correct answer equals that of a true or false question. Many times you will be able to determine the correct answer through elimination.

Never allow yourself to be influenced by an answer pattern.

If the last three or four answers were all the letter "d," there is no reason to believe that some pattern exists. Answer patterns are purely coincidental.

Careful reading of the question and the selection of answers cannot be overemphasized.

As an example, here is an actual question that has appeared on many exams and is usually answered incorrectly. Over 90% of all examinees will fail to select the correct answer.

    1. A rotating vane anemometer is an indicating instrument that registers an air current in:

       a. mph

       b. cfm

       c. fpm

       d. feet

Few people, even those who regularly use a rotating vane anemometer will select the correct answer: d. feet. The rotating vane anemometer measures feet of air, and the operator using the instrument uses a watch with a second hand to provide the time, usually clocked at one minute. The operator calculates the air speed in feet per minute but the instrument only indicates feet.

Do not rush into selecting an answer.

Simple questions with seemingly obvious answers are often missed because the examinee was simply in a hurry to move on to the next question when he or she was sure that the answer to a particular question was obvious. Read every word of the question and consider every single answer provided. Only then should you mark your answer sheet.

Study Tech Tip

Carefully read every question and the possible answers. Carefully eliminate the obviously incorrect answers and then select the correct answer.

Notes

Dealing with multiple-choice questions is a learned skill.
The more multiple-choice questions you deal with, the better you will develop your testing skills. The practice questions provided in this guide will help you develop that skill. Take advantage of the practice questions and use them as a method of study.

Finally, develop a "Test-Smart Attitude."
By practicing as much as possible, you will gain confidence in your test taking skills. The attitude you bring into the testing room will affect your ability to perform well. Be confident and build that confidence through practice.

PART THREE
About the Core Exam (includes practice exam)

What You Will Learn

- What a core exam is

- What the "soft studies" questions are

- The soft questions

Study Tech Tip

Leave the difficult questions for later. Answer the easier questions first. The easy questions carry the same percentage value as the difficult questions, so spend your time accumulating as many test points as possible. If you have time left you can return to the more difficult questions and raise your test score, but be very careful in lining up your answers to the appropriate question.

Not all of the testing organizations use a core exam, nor do they agree on what the core exam should contain. A core exam contains questions covering what the examining organization and their test writers believe are essential to all the specialties. Failing the core exam means that no credit will be given for any of the other exams, no matter how well you scored on them. The philosophy is that the material in the core exam is foundational to everything else. HVAC Excellence requires that everyone sitting for a specialty exam must first pass the 100-question multiple-choice electrical certification exam. HVAC Excellence believes that a solid understanding of the fundamentals of electrical systems as applied to air conditioning and heating is so important that it was made the core to all other exams in their series.

NATE takes a different approach and has developed a 50-question multiple-choice core exam covering a variety of topics, including grammar, vocabulary, customer relations, human resources, honesty, integrity and plain old common sense. These are the topics that many people have termed the "soft studies" questions to differentiate them from the math or science-type questions.

This section of the study guide will concentrate on the typical NATE Core exam with its softer questions. There are only 50 questions in the NATE Core exam, while there are 100 questions in each of the specialty exams. However, passing the core exam is essential to getting credit for passing the others. There are many versions of the core exam, with each containing a wide variety of questions from the NATE question bank. Since the scope of the core area is so wide, it is difficult to prepare for what could appear on the core examination. However, all is not lost. If you have graduated from high school, you should have the basic skills to successfully answer many of the questions. If you have taken a variety of general education classes in college, the core questions should be even easier.

THE SOFT QUESTIONS

English grammar and usage

The NATE examination committee believes that a strong grasp of the basics of the English language is important to the success of technicians. We communicate with customers, supervisors, wholesalers, the dispatcher and other technicians constantly. Misunderstandings in communication can cause a great deal of difficulty, lost time, lost sales, incorrectly performed work and even safety issues. Therefore, you may find several questions relating to the proper use of language. The test may ask questions about the parts of speech, including nouns, pronouns, adjectives, verbs, adverbs, conjunctions and prepositions. You may be asked to identify a poorly worded sentence from a series of sentences or find an error in punctuation. You will find examples of such questions in the 50-question core practice exam at the end of this section.

Spelling

The NATE examination committee recognizes how important spelling is to the proper performance of written communication in our industry. Misspelled words on an invoice explaining what service the technician provided can cause a customer to doubt the quality of the job rendered if the technician cannot spell. Many technicians are unable to properly spell the word "technician". Spelling questions are not uncommon on the NATE core examination.

Vocabulary, synonyms and antonyms

Once again, the exam writers know how important clear communication is to any workplace. Technicians need to know both commonly used words as well as technical terms and their meanings.

Reasoning ability

Technicians must have the ability to reason out solutions to problems, understand how physical relationships affect one another and how mechanical and electrical concepts apply to new situations. The core exam will contain general reasoning ability questions similar to those that appear on some mechanical aptitude tests.

Common knowledge

Technicians must have a solid grasp of basic math, units of measurement, time, basic geography and a number of facts everyone should know as basic survival skills. Some common knowledge questions may appear on the core exam.

Good judgment, honesty and integrity

The NATE examiners and the contractors they consulted decided the core should include questions that test the examinee for the use of good judgment and honesty, so several such questions will appear on the actual exam. Examples of such questions are included in the 50-question core practice exam in this section.

You are now presented with the first practice exam in this guide. The answers to the questions are in the back of the book. Do not look at the answers until after you have completed the entire 50-question core practice exam.

Study Tech Tip

Do not leave any questions unanswered. An unanswered question is counted against you the same as if you selected the wrong answer. Use the process of elimination and make an educated guess at the correct answer.

Study Tech Tip

Do not mark more than one answer to a question. Doing so will automatically score that question as incorrect. Carefully erase any answers you wish to change.

CORE PRACTICE EXAM

1. An air conditioning technician is working in the basement of a large public building. The technician notices several unsafe and dangerous conditions. Of the following conditions that are noticed, which is the most dangerous and should be reported immediately?

   a. Gas leaking from a cracked pipe

   b. An open sewer drain connection

   c. The basement is unlighted

   d. None of these are the technician's business

2. A service technician is returning home at 3 a.m. after completing an emergency service call. She notices heavy smoke coming out of the top floor of a building. Which of the following should she do?

   a. Make certain there is a fire

   b. Enter the building and warn the occupants

   c. Call the fire department

   d. Attempt to extinguish the fire before it gets out of control

3. An office worker frequently complains that his office is too cold or too warm. The best action to take is to:

   a. Ignore his complaints because they come from a habitual crank.

   b. Inform the office worker that the air conditioning system was designed by an engineer who failed miserably and you are unable to make a poorly designed system operate to meet the needs of the occupants.

   c. Investigate the problem and determine if an actual temperature control problem exists.

   d. Pretend to make an adjustment at the thermostat and inform the office worker that the problem has been solved.

4. A technician is taking evening classes in air balancing at the local college and needs photocopies of his project to distribute to others in the class this evening. The technician is on a service call in the office building of a long term customer and there are copy machines on every floor. Given the following options, what should the technician do?

    a. Find a copy machine that is not occupied, is in a remote area of the building and quickly make his copies.

    b. Locate someone who knows him and ask if he can make copies of his project.

    c. Pretend he is making copies of paperwork related to his work for that customer and just make the copies.

    d. Make his copies and leave some money on the machine.

5. While you are working at a construction site an electrician working nearby asks you to help him move a pallet of electrical materials. Of the following, the best action for you to take is:

    a. Inform the electrician that moving electrical materials is not the job of a sheet metal installer and he should do it himself or find another electrician.

    b. Help the electrician move the pallet and return to your own work.

    c. Tell the electrician you will help him after you complete your own work.

    d. Tell him your work is as important to you as his work is important to him, and continue with your own work.

Study Tech Tip

Do not try to memorize any of the practice questions and their answers. Learn the material or concept behind the questions to know why the correct answers are correct and why the incorrect answers are incorrect. The real exam questions are changed on a regular basis or are reworded.

Study Tech Tip

One study claims that the average American spends four hours a day watching television. Imagine what could be accomplished if you spent only half that time each day reading or studying in your chosen profession!

6. A large crane is setting a heavy packaged air conditioning system on the roof of a building. You are assigned the job of preventing all traffic from entering the area where the crane is operating. You have been given strict orders not to allow any vehicle to enter the area while the work is in progress. You block the alley with your service van and remain in the driver's seat waiting for the "all clear" signal. Suddenly, an ambulance arrives with its lights and siren on and obviously wants to enter the alley you are blocking. Given the following, what is the best action to take?

   a. Get out of your van and calmly explain that you are not to allow any vehicle to pass until the crane has finished the lift.

   b. Move your service van and allow the ambulance to pass.

   c. Get permission from the crane supervisor or operator before allowing the ambulance through.

   d. Inform the ambulance driver that if you let him pass you will lose your job.

7. The square root of 81 is:

   a. 81

   b. 9

   c. 1

   d. 8

8. The square of 4 is:

   a. 16

   b. 9

   c. 2

   d. 8

9. How many square inches are in a square foot?

    a. 144

    b. 100

    c. 12

    d. 3.14

10. How many hours are in a year?

    a. 5,280

    b. 6,000

    c. 8,760

    d. 10,000

11. What is the freezing temperature of water?

    a. 30 degrees

    b. 32 degrees

    c. 0 degrees F

    d. 144 btu/lb

12. A 6-inch pulley rotating at 100 rpm drives a larger 8-inch pulley. The 8-inch pulley is:

    a. Rotating at less than 100 rpm

    b. Rotating faster than 100 rpm

    c. Rotating at the same speed as the 6-inch pulley

    d. Rotating at twice the speed as the 6-inch pulley

Study Tech Tip

Never allow yourself to be influenced by an answer pattern. If the last several questions had an answer of "a", there is no reason to think the next answer will be the same or different. There is never a pattern to the answer key.

Study Tech Tip

Do not rush to answer a question because you think you absolutely know the answer. Read each of the possible answers carefully. Several could be very close to being correct but only one is the actual correct answer.

13. A gallon of water weighs:

    a. 1 pound

    b. 1.5 pounds

    c. 8.33 pounds

    d. 20 pounds

14. The word "compound" means:

    a. Multiple

    b. Single

    c. Atomic

    d. Difficult

15. A ruler is:

    a. Digital

    b. Analog

    c. Linear

    d. Non-linear

16. The word "sensor" means:

    a. Response

    b. Detector

    c. Inner-mind

    d. To remove

17. The word "renovate" means:

    a. To change

    b. Imitate

    c. Restore

    d. Cover

18. The outside perimeter of a circle is called the:

    a. Diameter

    b. Radius

    c. Circumference

    d. Arc

19. A ton is how many pounds?

    a. 12,000

    b. 2,000

    c. 1,500

    d. 1,000

20. Two eighths is the same as:

    a. One fourth

    b. One half

    c. Three quarters

    d. One inch

Study Tech Tip

Develop a "Test Smart Attitude." Practice your test-taking skills and gain confidence with reading, understanding and choosing the correct answer from the choices available.

Study Tech Tip

Learning new material can be more interesting, more fun and easier to comprehend if you spend time discussing and reviewing the material with another technician. However, make sure you spend time studying by yourself as well.

21. The word "phase" means:

    a. Trouble

    b. Electricity

    c. Change

    d. Shadow

22. "RAM" stands for:

    a. Remote access motion

    b. Random access motion

    c. Random access memory

    d. Relay activated motion

23. A "hard drive" is used to:

    a. Store and retrieve information

    b. Input information

    c. View information

    d. Calculate numbers

24. A "CPU" is used to:

    a. Store and retrieve information

    b. Sort and calculate

    c. View information

    d. Recall information

25. The 24th letter of the alphabet is:

    a. W

    b. Z

    c. Y

    d. X

26. The best reason for overhauling a machine often is:

    a. Overhauling is easier to do when done often

    b. To minimize breakdowns of the machine

    c. To make sure the machine is lubricated

    d. To make sure we understand how the machine operates

27. The best method to put out a gasoline fire is:

    a. Use a bucket of water

    b. Smother it with rags

    c. Use a carbon dioxide fire extinguisher

    d. Use a carbon tetrachloride fire extinguisher

28. If it takes 4 days for three machines to do a certain job, it will take two machines:

    a. 6 days

    b. 5.5 days

    c. 5 days

    d. 4.5 days

Study Tech Tip

Information that is organized into logical groups and sequences is learned more easily. Studying new material within the framework of its topic increases retention.

That is one of the reasons this book has grouped practice questions by topic such as electrical, gas heating, heat pumps and others.

29. "Plumber" is related to "Wrench" as "Painter" is related to:

   a. Brush

   b. Pipe

   c. Shop

   d. Hammer

30. "Research" is related to "Findings" as "Training" is related to:

   a. Skill

   b. Tests

   c. Supervision

   d. Teaching

31. Which of the following is the correct spelling?

   a. Anticipate

   b. Antisipate

   c. Anticapate

   d. None of these

32. Which of the following is the correct spelling?

   a. Condenser

   b. Condensor

   c. Condinsor

   d. Condinser

33. Which of the following is the correct spelling?

    a. Similar

    b. Simmilar

    c. Similiar

    d. None of these

34. Which of the following is the correct spelling?

    a. Compresser

    b. Compressor

    c. Compreser

    d. Compreser

35. Which of the following is the correct spelling?

    a. Technician

    b. Techniton

    c. Tecknition

    d. Techniction

36. Frequent use of marijuana may impair your judgment. Impair means:

    a. Weaken

    b. Conceal

    c. Improve

    d. Expose

Study Tech Tip

In a survey, 73 percent of contractors said having certified technicians was integral to their business plan and success as a contractor.

37. Analyze the underlined portion of the following sentence and choose a correction from the answers provided. "There <u>weren't no</u> filters in the van."

   a. The sentence is correct as is

   b. weren't any

   c. wasn't any

   d. wasn't no

38. Heat flows from an object or area at a higher temperature to an object or area at a lower temperature. Therefore, your furnace in your home:

   a. Heats you

   b. Slows down your heat loss

   c. Humidifies you

   d. None of these

39. The word "micro" means:

   a. Opaque

   b. Small

   c. Complete

   d. Miniature

40. The word "random" means:

   a. With space

   b. Without order

   c. Slow moving

   d. Changing

41. The word "orbital" means:

    a. Enclosing

    b. Circling

    c. Projecting

    d. Loud

42. The word "accept" means:

    a. To receive

    b. To omit

    c. To exclude

    d. None of these

43. The word "affect" means:

    a. To influence

    b. A result

    c. To bring about

    d. None of these

44. The word "amount" means:

    a. How many you have

    b. How much you have

    c. Both a and b are acceptable

    d. None of these

45. To "compare" means:

    a. To point out similarities

    b. To point out differences

    c. Both a and b are acceptable

    d. None of these

Study Tech Tip

"Any technician can claim to be good but the certified technician can prove his or her worth"

46. Which of the sentences below is correctly punctuated?

   a. His room was littered with books, pens, paper, and maps.

   b. His room was littered, with books, pens, paper, and maps.

   c. His room was littered with books, pens, paper and maps.

   d. His room, was littered with, books, pens, paper, and maps.

47. Which of the following sentences is grammatically incorrect?

   a. One can easily spot your mistakes if you check carefully.

   b. One can easily spot one's mistakes if you check carefully.

   c. You can easily spot your mistakes if you check carefully.

   d. All three are grammatically correct.

48. Which of the following sentences is grammatically incorrect?

   a. John is the tallest of the two brothers.

   b. John is the taller of the two brothers.

   c. Neither a nor b is correct.

   d. Both a and b are correct.

49. A verb:

   a. Names a person, place or thing.

   b. Stands in the place of a noun to avoid repeating it.

   c. Expresses action

   d. Joins or shows the relationship between words, phrases or clauses.

50. A conjunction:

   a. Names a person, place or thing

   b. Stands in the place of a noun to avoid repeating it

   c. Expresses action

   d. Joins or shows the relationship between words, phrases or clauses

PART FOUR
Math Practice Exam

What You Will Learn

- What types of math questions and problems to expect on the exams

- How to deal with a variety of math questions

- As you take and review the practice exams you will gain confidence in your math ability

Math Tech Tip

To determine the temperature

difference between two

temperatures where one of

the temperatures is below

zero, add the two numbers

For example, the difference

in temperature between 60

degrees above zero and 10

degrees below zero is found by

adding the two. Sixty plus ten

equals 70 degrees difference.

Technicians deal with mathematics on a daily basis. The complexity of the math varies widely depending upon the specific duties of the technician, his or her area of specialization, and the complexity of the technician's work. Every technician certification exam contains math problems and questions regarding mathematical relationships.

This section of the guide contains a practice math exam designed to help you tackle and successfully solve the wide variety of math problems that occur on the exams. There is no specific math section on the actual exams . The math questions may randomly occur anywhere and apply directly to actual problems a technician may need to solve on the job.

The following 50-question practice exam provides a good sampling of the type of math questions often found on technician certification exams. Typically, the NATE exams will each contain approximately 6 to 12 of these types of math questions. The NATE core exam may also contain several math or mathematical reasoning questions. The RSES CM or Certificate Member exam may contain more math-related problems. The actual exams will mix the math questions with other technical, code, safety and practical questions.

Before attempting to actually calculate the answers, carefully examine the questions and the selection of answers available. Occasionally, the correct answer is obvious because the distracters (false answers) are not logically correct.

After taking and correcting this practice exam, go back over those you missed and determine what you did incorrectly. Also, practice those types of math problems until you are comfortable you can solve any of that type.

MATH AND MATHEMATICAL REASONING PRACTICE EXAM

Math Tech Tip

There are 144 square inches in a square foot. Therefore, a duct opening measuring 8 inches by 24 inches equals a duct opening area of 192 square inches when multiplied. Since there are 144 square inches in a square foot, 192 square inches divided by 144 square inches equals 1.33 square feet of area. Converting square inches to square feet is often necessary when working with duct systems.

1. A heating system has an input of 100,000 btuh with an output of 75,000 btuh. What is its efficiency in percent?

   a. 100%

   b. 80%

   c. 75%

   d. 70%

2. A motor has a power factor of .69, a voltage rating of 230 volts, 3-phase and an efficiency rating of 88% with an SF rating of 1.15. Which of these ratings relates to the phase angle between the current and the voltage?

   a. Power factor

   b. Voltage

   c. Phase

   d. SF

3. The decimal equivalent of the mixed fraction $3\frac{7}{8}$ is:

   a. 3.125

   b. 3.375

   c. 3.750

   d. 3.875

4. The number on the top of a fraction is called:

   a. Quotient

   b. Numerator

   c. Denominator

   d. Divisor

Math Tech Tip

Compression Ratio must be calculated by changing psig to psia in order to make the ratio mathematically correct. For example, if a system has a suction pressure of 10 psig and a high side pressure of 100 psig the compression ratio would be 10:1 if not properly converted to psia as required. The correct ratio is 5:1 after adding 14.7 psi to the suction and discharge pressures.

5. The number on the bottom of a fraction is called:

    a. Quotient

    b. Numerator

    c. Denominator

    d. Divisor

6. Air balancing reports generally round off all cfm air flow rates to the nearest 5 cfm. An airflow measurement of 348.3 cfm will then be rounded off to which of the following?

    a. 340 cfm

    b. 345 cfm

    c. 348 cfm

    d. 350 cfm

7. A technician is using a U-tube manometer to measure a pressure differential. One side of the U-tube water level is at +.75". The opposite side of the U-tube reads −.75". What is the pressure differential in inches?

    a. .75"

    b. 1.25"

    c. 1.5"

    d. 1.75"

8. A technician calculates the bill for a service call to be a total of $269.60. The sales tax rate is 5%. How much is the sales tax for the bill?

    a. $13.48

    b. $14.67

    c. $26.96

    d. $283.08

9. The velocity in fpm is equal to the square root of the velocity pressure times the factor 4005. A technician measures a velocity pressure of .25. What is the square root of .25?

   a. .5

   b. .05

   c. .025

   d. .625

10. A technician has used 50 feet of four conductor thermostat wire from a 500 foot roll. What is the value of the remainder of the wire on the roll if the cost of the wire is 10 cents per foot?

    a. $25

    b. $45

    c. $50

    d. $55

11. A technician pulls 156 inches of thermostat wire through a crawl space. How many feet of thermostat wire did he pull?

    a. 15.5 feet

    b. 13 feet

    c. 18 feet

    d. 18.5 feet

12. A technician clocked a furnace as using 40 cubic feet of natural gas in 60 minutes. How many cubic feet did the furnace use per minute?

    a. .5 cubic feet per minute

    b. 5.5 cubic feet per minute

    c. .66 cubic feet per minute

    d. 6.6 cubic feet per minute

Math Tech Tip

The circumference or perimeter of a circle is the distance around the outside of the circle. The diameter is the distance across the circle at the center, and the radius is one half the diameter of the circle or the distance from the center to the outside edge.

13. A motor operating a blower has a belt drive system. The motor pulley is 6 inches in diameter and operates at 1800 rpm. The blower pulley diameter is 12 inches. What is the rpm of the blower?

    a. 3600 rpm

    b. 1800 rpm

    c.   900 rpm

    d.   800 rpm

14. A 120 to 24 volt step down control transformer has a VA rating of 48 watts. What is the maximum amperage the transformer should be allowed to draw on the 24 volt side?

    a. 2 amps

    b. 2.5 amps

    c. 5 amps

    d. 5.5 amps

15. A technician needs to calculate the compression ratio for a system operating with a high side pressure of 200 psig. In order to compute the ratio, the pressure must first be converted to psia. What is the psia equivalent for 200 psig?

    a. 185 psia

    b. 215 psia

    c. 230 psia

    d. 300 psia

16. The internal dimensions of a building are 10 feet high, 100 feet long and 50 feet wide. How many cubic feet of air are contained in the building?

    a.   500 cubic feet

    b.   750 cubic feet

    c.   5000 cubic feet

    d. 50,000 cubic feet

17. Using the internal volume from the previous question, if the density of air is .075 lbs per cubic foot, what is the weight of the air in the building?

    a.   37.5 lbs

    b.   375. lbs

    c.   3,750 lbs

    d. 37,750 lbs

18. A technician purchases a gallon of POE oil for $120. The technician uses one quart of the oil on a job. The profit mark-up to be applied is 2 times the cost. How much is the marked up cost of the quart of oil?

    a. $ 15.00

    b. $ 30.00

    c. $ 60.00

    d. $120.00

19. A furnace manufacturer is offering a 20% discount off their normal wholesale price to dealers. If the normal wholesale price is $400.00 what is the discounted price for the furnace?

    a. $300.00

    b. $320.00

    c. $325.00

    d. $380.00

Math Tech Tip

A three-phase motor produces 73% more horsepower than a single-phase motor at the same voltage and current draw. The 1.73 multiplier used in three-phase calculations takes this fact into consideration. The square root of 3 is 1.73

Electrical Tech Tip

Increasing the airflow of a blower by even a small amount increases the load on the blower motor substantially and can cause the motor to overheat or burn out. The horsepower required increases by the cube of the increase in rpm or cfm. Do not over-amp a motor!

20. What is the volume of a cylindrical water tank that is 6 feet in diameter and 10 feet tall?

    a. 826.2 cubic feet

    b. 826.2 cubic inches

    c. 600 cubic inches

    d. 282.6 cubic feet

21. Water weighs 8.33 lbs per gallon, there are 7.5 gallons per cubic foot and one cubic foot of water weighs 62.4 pounds. What is the weight of the water in the tank in problem 20? Assume the tank is full.

    a. 17,634.24 lbs

    b. 16,746 lbs

    c. 8,333 lbs

    d. 62,400 lbs

22. How many gallons of water can the tank in problems 20 and 21 hold?

    a. 2,116.95 gallons

    b. 2,351.23 gallons

    c. 833.333 gallons

    d. None of these is correct

23. There are 746 watts per horsepower. How many horsepower are there per watt?

    a.   .746

    b.   7.46

    c. 74.6

    d. none of these

24. One ton of cooling is at a rate of 12,000 btuh. How many btu per day is this? Use a 24-hour day as a normal day.

    a. 200

    b. 200,000

    c. 288,000

    d. 300,000

25. Convert minus 40 degrees Fahrenheit to Centigrade:

    a. +40

    b. +20

    c. -20

    d. -40

26. How many degrees are in a quarter of a circle?

    a. 180

    b. 90

    c. 60

    d. 45

27. A duct measures 24" high by 36" wide. How many square feet of duct area are there?

    a. 3 square feet

    b. 4 square feet

    c. 6 square feet

    d. 8 square feet

Motor Tech Tip

When replacing a motor with a more efficient motor be aware that many energy efficient motors may have lower amperage ratings for the same horsepower.

Motor Tech Tip

Vibration from motors can

cause noise to be transmitted

far from its source. Resonance

is sympathetic vibration that

takes place in objects that are

not the source of the vibration

but which are prone to vibrate

at the same frequency or a

multiple of the frequency

generated by the source.

Sometimes the problem can

be solved by making a small

change in the speed of the

motor thus changing the

frequency which may eliminate

the resonant vibrations.

28. The U value of a wall is the reciprocal of the R value. A wall with an R value of 10 has just had enough insulation added to increase the R value by 5. What is the new U value of the wall to two decimal places?

   a. 15.

   b. 5.

   c. .66

   d. .06

29. The volume of a theatre is 100,000 cubic feet. Standard air has a volume of 13.33 cubic feet per pound and a density of .075 lbs per cubic foot. What is the weight of the air in the theatre?

   a.   750 lbs

   b.   133.3 lbs

   c.   7,500 lbs

   d.   13,300 lbs

30. The specific heat of air is .24 btu/lb/degree. The density of air is .075 lbs/cubic foot. In the equation, Btuh = 1.08 × CFM × TD, where does the factor 1.08 come from?

   a. It converts minutes to hours, cubic feet to pounds and includes the specific heat of air.

   b. It converts minutes to hours, pounds to cubic feet and specific heat to temperature difference.

   c. It converts cubic feet to cubic inches, specific heat to temperature difference and hours to minutes.

   d. It is a number used to average out the values and give an approximate answer for the equation.

31. How much heat in btu is required to raise the temperature of a full 30 gallons of water in a water heater from 50 degrees to 100 degrees F?

    a. 12,495

    b. 13,680

    c. 15,000

    d. 20,000

32. A 5-ton air conditioning system has a btuh capacity of:

    a. 40,000

    b. 45,000

    c. 50,000

    d. 60,000

33. Assuming a normal comfort cooling application, the average cfm per ton is 400. How many cfm should a 5-ton comfort cooling air conditioning system move through its air system if the system utilizes a 7.5 hp motor at 230 volts, 3-phase and uses a centrifugal blower with a belt drive?

    a. 1,000 cfm

    b. 2,000 cfm

    c. 2,500 cfm

    d. 3,000 cfm

34. Electric resistance heaters give off 3.42 btu per watt. How much heat in btuh will an electric resistance space heater give off if it draws 20 amps at 230 volts?

    a. 12,732

    b. 15,732

    c. 16,000

    d. 18,230

Safety Tech Tip

Never stand in front of a motor starter, contactor or capacitor when starting equipment while performing service with the control panel open.

Safety Tech Tip

A fuse is designed to protect equipment and systems. A GFCI (Ground Fault Circuit Interrupter) is designed to protect people. In any event never rely on any device to protect against electric hazards. Learn and follow all electrical safety procedures when working on or around electrical systems.

35. Acme Heating's new service van used 18 gallons of fuel and traveled 200 miles. What is the gas mileage for the new van?

   a. 8 mpg

   b. 10 mpg

   c. 11 mpg

   d. 12 mpg

36. It took two installers 10 hours total to install a new air conditioning system in the first of several homes going into a new tract. Every home is identical so the rest of the installations will be identical. If the two installers can reduce their installation time by 10%, how many hours should it take them per install?

   a. 6 hours

   b. 7 hours

   c. 8 hours

   d. 9 hours

37. The saturation temperature of an evaporator is 40 degrees F. The evaporator uses an external equalized thermostatic expansion valve. The temperature measured at the sensing bulb is 50 degrees F, and the temperature measured 20' away and 6 inches from the compressor's suction service valve is 65° F. What is the superheat of the thermostatic expansion valve?

   a. 10 degrees

   b. 15 degrees

   c. 20 degrees

   d. 50 degrees

38. A compressor operates with a high side pressure of 100 psig and a low side pressure of 10 psig. What is the compression ratio for the compressor?

    a. 10:1

    b. 8:1

    c. 4.6:1

    d. 2.5:1

39. An evaporator has 1600 CFM of airflow across it and is removing 30,000 btuh of heat. What is the temperature drop of the air across the evaporator?

    a. 17.36

    b. 18.75

    c. 19.25

    d. 20.50

40. A hot water heating coil has an actual output of 40,000 btuh and has a water temperature differential of 10 degrees between the supply and return pipes. What is the water flow rate in gpm through the coil?

    a. 6 gpm

    b. 8 gpm

    c. 10 gpm

    d. 12 gpm

Motor Tech Tip

Never carry an electric motor

by its leads. It can easily

be damaged.

41. A technician measures an air velocity of 400 fpm through a duct that measures 24-inches square. What is the airflow quantity in CFM?

     a.   400 cfm

     b.   600 cfm

     c.   800 cfm

     d.   1,600 cfm

42. Code restricts filling refrigerant cylinders to more than 80% of their total capacity. How many pounds of refrigerant are allowed in a 150 pound capacity cylinder?

     a.   100 lbs

     b.   120 lbs

     c.   125 lbs

     d.   130 lbs

43. How many pounds of water can be heated by 6 degrees with the addition of 72 btu?

     a.   .833

     b.   12

     c.   432

     d.   72

44. What is the size of a supply duct that passes 800 cfm of air at 600 fpm velocity?

     a.   133 square inches

     b.   192 square inches

     c.   239 square inches

     d.   414 square inches

45. A clamp-on ammeter reads 20 amperes on a purely resistive circuit and the voltage is 120 volts. What is the wattage in kW?

 a. 2 kW

 b. 2.4 kW

 c. 24 kW

 d. 240 kW

46. A typical frame wall is constructed of the following materials where the individual R values for each material is known.

i. Outside air film	R = .17
ii. Wood siding	R = .85
iii. Sheathing	R = 2.06
iv. Air space	R = .97
v. Plaster board	R = .41
vi. Inside air film	R = .68

The U value for this wall is:

 a. 5.14

 b. 4.27

 c. .27

 d. .19

47. A boiler with an input of 120,000 btuh and which has an output of 102,000 btuh has a numerical efficiency of:

 a. 75%

 b. 80%

 c. 85%

 d. 90%

48. A duct that has a circumference of 31.416" has a diameter of:

    a. 10"

    b. 11"

    c. 12"

    d. 16.5"

49. Increasing a duct diameter from 4 inches to 8 inches will increase its area by:

    a. 2 times

    b. 2.5 times

    c. 3 times

    d. 4 times

50. The square root of 25 is:

    a. 8

    b. 5

    c. 2.5

    d. 12

PART FIVE
Electrical Certification and Practice Exam

What You Will Learn

- What to expect from the electrical portion of the certification exams

- Why the certification exams contain so many electrical questions

- Confidence and experience as you take and review the electrical practice questions

Electrical Tech Tip

A switch that closes (energizes) on an increase in temperature, pressure, humidity or other variable is called a direct acting switch. A switch that opens (de-energizes) upon an increase in temperature, pressure, humidity or other variable is called a reverse acting switch. A control containing a switch that provides both functions is called a dual acting control.

The topic of electricity and applications to HVAC electrical systems is a very important area for the technician to master. For that reason, HVAC Excellence has determined that passing their 100-question electrical exam is required prior to obtaining any of their certifications. The first time you sit for an HVAC Excellence exam, you take their 100-question electrical exam plus one of their other 100-question specialty exams. Additional certifications may be earned without retaking the electrical exam. Making an electrical exam as their core exam makes a great deal of sense. No amount of electrical knowledge is ever enough! Some HVAC authorities claim that more than 80% of a service technician's work is electrical.

Electrical questions are so prevalent they will occur in every certification exam you take. Spending extra study time on electricity, electronics, electrical math and reading wiring diagrams not only pays off on testing day, but it is also highly useful on the job. Remember, electrical math problems will also appear with the electrical theory and practical questions. The following practice electrical exam contains questions much like those that will appear in any of the electrical certification exams.

ELECTRICAL PRACTICE EXAM

1. What electro-mechanical control would you use on a cap tube or automatic expansion valve system to maintain the space temperature?

   a. A dual pressure control

   b. A low pressure control

   c. A thermostatic control

   d. A high pressure control

2. Open motors are designed to operate at a given temperature rise as indicated on the motor nameplate. What does this mean?

   a. It indicates the temperature rise over the surrounding air of a motor when the motor is operating at full-load conditions.

   b. It indicates the temperature rise of the motor.

   c. It indicates the temperature drop of the surrounding air.

   d. It indicates the temperature drop of the motor at rest.

3. A voltmeter is connected into a circuit in which of the following ways?

   a. Series

   b. Parallel

   c. Series-parallel

   d. Delta

4. A low pressure control is usually located on the:

   a. Condensing unit

   b. Evaporator

   c. Cabinet

   d. Blower

Electrical Tech Tip

The VA rating on a control transformer is called the wattage rating. Dividing the VA rating by the output voltage rating of the secondary gives the maximum amperage the secondary winding can provide before burning up the transformer.

Safety Tech Tip

Stay clear of power lines!

The minimum clearance to a power line for your safety is 4-feet. For power lines over 50,000 volts maintain a clearance of 10 feet or more. Watch where you place your ladder when accessing a roof. Power lines are nearly always nearby.

5. A high pressure control is usually located on the:

    a. Condensing unit

    b. Evaporator

    c. Cabinet

    d. Blower

6. The term "nominal voltage" refers to:

    a. The actual voltage

    b. The measured voltage

    c. The average voltage

    d. The listed voltage

7. Which of the following is a safety control?

    a. Thermostat

    b. High pressure control

    c. Low pressure control

    d. Humidistat

8. Low voltage controls are usually designed to operate at no more than:

    a. 25 volts

    b. 80 volts

    c. 120 volts

    d. 230 volts

9. A unit of power equal to one horsepower and delivered at the shaft of a motor is known as:

    a. Amperage

    b. Brake horsepower

    c. Horsepower

    d. Voltage

10. What is the purpose of the holding circuit in a magnetic motor starter?

    a. Holds the main contacts open until the control circuit through the interlock is made.

    b. Holds the main contacts closed until the control circuit through the interlock is broken.

    c. Holds the main contacts open until the control circuit through the interlock is broken.

    d. Holds the main contacts closed until the control circuit through the interlock is made.

11. On a magnetic motor starter, the alloy piece that causes the motor starter to open and shut off the motor if too much motor current is drawn is called the:

    a. Holding coil

    b. Heater

    c. Relay

    d. Trip armature

12. A set of contacts in a pressure control that close when the pressure increases are:

    a. Direct acting

    b. Reverse acting

    c. Dual acting

    d. None of these

13. The bimetal strip in a common non-programmable thermostat is the:

    a. Anticipator bimetal

    b. The differential bimetal

    c. Contactor bimetal

    d. None of these

Electrical Tech Tip

Always wind a wire around its screw terminal in the same direction the screw turns when tightened so that the wire is tightened with the screw.

Electrical Tech Tip

Ground fault interrupters (GFI) are highly desirable electrical safety devices that continually monitor the currents between the hot wire and the neutral wires. The currents should be perfectly equal unless current is bleeding off somewhere it does not belong, such as to ground, the enclosure of an appliance, a power tool or machine. The GFI instantly trips the circuit if the difference in currents is .005 amps or greater.

14. Which of the following control settings is typically non-adjustable?

    a. Set point

    b. Range

    c. Differential

    d. Cut-in

15. Ohm's law states:

    a. $E = IR$

    b. $E = PR$

    c. $R = IE$

    d. $I = RP$

16. All else being equal, which of the following wire sizes will carry the most current?

    a. 10 ga

    b. 14 ga

    c. 12 ga

    d. 18 ga

17. The run capacitor on a motor is used to:

    a. Increase torque

    b. Increase horsepower

    c. Decrease resistance

    d. Increase amperage

18. Doubling the voltage to a resistance will cause the power or watts to:

    a. Double

    b. Triple

    c. Quadruple

    d. Remain the same

19. What is impedance?

    a. Opposition to alternating current flow

    b. Electrical flow

    c. Amperage

    d. Electrical pressure

20. How are fuses sized?

    a. By amperage

    b. By microfarads

    c. By voltage

    d. Both a and c

21. Capacitors are rated in:

    a. Amperes

    b. Microfarads

    c. Volts

    d. Both b and c

22. One horsepower is equal to _____ watts.

    a. 3.414

    b. 746

    c. 1.08

    d. 1200

23. How much resistance must be placed in a 220-volt circuit to limit the current to 2 amps?

    a. 55 ohms

    b. 110 ohms

    c. 30 ohms

    d. None of the above

Electrical Tech Tip

A device that increases in resistance with an increase in temperature is said to have a positive temperature coefficient, or PTC. The higher the coefficient, the greater the increase in resistance per degree of temperature rise. Solid-state compressor starting relays are usually PTC devices with a high coefficient. Conversely, a material with a negative temperature coefficient, or NTC, drops in resistance when heated. Most conductors are PTC materials, while some semiconductors are NTC materials. PTC and NTC electronic sensors are used in electronic devices applied to heating and cooling systems.

Troubleshooting Tech Tip

Do not turn off the power of a system that uses impedance relays, lockout relays or reset relays until you have determined which control device or devices may have locked out the circuit. This will help guide you to the cause of the problem.

24. Impedance in a circuit is the result of:

    a. Pure resistance

    b. Inductive reactance

    c. Capacitive reactance

    d. All of the above

25. Doubling the voltage to a fixed resistance will:

    a. Not change the current

    b. Cause the current to double

    c. Cause the current to triple

    d. Cause the current to increase by a factor of four

26. When the temperature of a negative temperature coefficient material increases, its resistance will:

    a. Remain the same

    b. Increase

    c. Decrease

    d. Change by the square of the applied voltage

27. What is the Btuh output of an electric heater rated at 15 Kw?

    a. 51.15 Btuh

    b. 51,150 Btuh

    c. 5,115 Btuh

    d. None of the above

28. What is the total capacitance of two 20-microfarad capacitors wired in series with each other?

    a. 5 mf

    b. 10 mf

    c. 20 mf

    d. 40 mf

29. What is the total capacitance of two 20-microfarad capacitors wired in parallel with each other?

    a. 5 mf

    b. 10 mf

    c. 20 mf

    d. 40 mf

30. If a conductor is wrapped around the jaws of a clamp-on ammeter four times and the meter indicated 16 amps, the actual amperage would be:

    a. 16 amps

    b. 8 amps

    c. 4 amps

    d. 2 amps

31. On a standard electro-mechanical thermostat, if the fan switch is in the "on" position, the indoor fan will:

    a. Run only when the thermostat is calling for the compressor to run

    b. Run when the indoor temperature reaches set point

    c. Run continuously

    d. None of the above

32. Which of the following capacitors is the best choice to replace a 35 mf, 370 volt run capacitor?

    a. 35 mf, 330 volt

    b. 35 mf, 390 volt

    c. 30 mf, 440 volt

    d. 40 mf, 370 volt

Troubleshooting Tech Tip

Voltage across a switch means the switch is open, not necessarily that the switch is bad. Determine what could cause the switch to open. For example, an open high-pressure control could be caused by a lack of airflow over the condenser coil, a refrigerant overcharge or air in the system. An open furnace draft switch could be caused by an obstruction in the flue.

Safety Tech Tip

Never assume the power is shut off to the system you are servicing. Many systems are powered from more than one source. Use lock-out and tag-out devices on every job and double check that the power is actually off by using a quality voltmeter. Make sure your meter is working by checking an energized circuit.

33. A motor is a _____ load.

   a. Resistive

   b. Inductive

   c. Capacitive

   d. Mechanical

34. The compressor oil safety control is an electro-mechanical control that operates on which two pressures?

   a. Oil pump pressure and suction pressure

   b. Oil pump pressure and discharge pressure

   c. Suction pressure and discharge pressure

   d. Oil pressure and atmospheric pressure

35. What is the millivolt reading of a good thermocouple?

   a.　5 mv

   b.　15 mv

   c.　30 mf

   d. 120 mv

36. A high pressure safety control is:

   a. Direct acting

   b. Reverse acting

   c. Dual acting

   d. None of the above

37. Two silicon-controlled rectifiers wired in reverse parallel with each other allows:

    a. Current flow in both directions

    b. Current flow from the anode to the gate

    c. Current flow from the cathode to the gate

    d. Current flow in one direction only

38. Hot surface igniters operate on which of the following voltages?

    a. 24 volts

    b. 120 volts

    c. 240 volts

    d. about 8 to 10 Kv

39. Which of the following mathematical relationships is true of a motor's horsepower if the CFM of a blower were changed?

    a. Hp changes directly proportional to changes in CFM

    b. Hp changes inversely proportional to changes in CFM

    c. Hp changes by the square of the change in CFM

    d. Hp changes by the cube of the change in CFM

40. When four diodes are constructed in a bridge circuit and alternating current is brought into the input of the bridge, the output of the bridge provides:

    a. Half wave rectification

    b. Full wave rectification

    c. Pulsating ac

    d. None of the above

Electrical Tech Tip

Doubling the voltage to a fixed resistance will cause the wattage (power) to increase by a factor of four. This is because the wattage is the current times the voltage, and doubling the voltage doubles the current. Twice the voltage times twice the current produces four times the wattage.

Electrical Tech Tip

There are only three kinds of electrical loads: Resistive loads, inductive loads and capacitive loads. Ohm's law only works correctly with resistive loads without correcting for power factor and efficiency.

41. When photons of light strike the surface of a typical cadmium sulfide photoconductive cell, _____.

    a. resistance increases

    b. resistance decreases

    c. cell produces voltage

    d. None of the above

42. The coil of a lockout relay is wired in _____ with the contactor coil and in _____ with the contacts of the control it is protecting.

    a. series, series

    b. parallel, parallel

    c. series, parallel

    d. parallel, series

43. A low voltage control system that primarily operates on two levels of voltage is called:

    a. An analog system

    b. A digital system

    c. An alternating system

    d. A DC system

44. A voltage imbalance is a problem associated with which type of motor?

    a. Single phase motors

    b. Three phase motors

    c. DC motors

    d. PSC motors

45. An air conditioning compressor has the following motor protection devices: low pressure control, high-pressure control, internal thermostatic overloads and an oil failure control. If the internal thermostatic overloads open, which of the following will take place?

    a. High pressure control will open

    b. Low pressure control will open

    c. Oil failure control will trip off and lock out the compressor

    d. After the motor cools the overloads will reset and the compressor will restart

46. Using the standard resistor color code to determine the value of a resistor, what is the resistance of a resistor with the following three colors in this order: red, green, black?

    a. 25 ohms

    b. 250 ohms

    c. 360 ohms

    d. None of the above

47. What is the purpose of the resistor wired across the terminals of a start capacitor?

    a. For the safety of the technician

    b. To prevent arcing and possible damage to the start relay contacts

    c. To increase the life of the start capacitor

    d. To protect the capacitor from back-EMF

48. Which of the following will cause an increase in the horsepower required per ton of cooling on an air conditioning system?

    a. Increasing discharge pressure

    b. Decreasing suction pressure

    c. Both a and b

    d. None of the above

Electrical Tech Tip

Inductive reactance is resistance in a circuit due to magnetic field effects, while capacitive reactance is resistance in a circuit due to the action of capacitance. Pure resistance is caused by the type of material, cross-sectional area, length and temperature. Impedance is the term for the total resistance in a circuit due to all forms of resistance combined.

49. Using an automatic pump down system, the compressor is directly controlled by:

    a. A thermostat

    b. Low pressure control

    c. High pressure control

    d. Oil pressure control

50. Which of the following test results indicate a capacitor is good?

    a. An ohmmeter is set to the highest ohm scale and its test leads placed across the capacitor terminals. The analog meter's needle moves quickly to the right side of the scale and then slowly returns to the left end of the scale.

    b. An ohmmeter is set to the lowest ohm scale and its test leads placed across the capacitor terminals. The analog meter's needle moves slowly to the right side of the scale and then drops suddenly back to the left end of the scale.

    c. An ohmmeter is set to the highest ohm scale and its test leads placed across the capacitor terminals briefly. When one of the test leads is removed a spark is observed.

    d. A voltmeter is connected across the terminals of a capacitor and no voltage is measured.

PART SIX
Air Conditioning Certification and Practice Exams

What You Will Learn

▪ What to expect from the air conditioning certification exams

▪ Confidence and experience as you take and review the air conditioning practice questions

Instrument Tech Tip

With an instrument that uses liquid in a transparent glass or plastic tube to take a reading, the liquid clings to the sides of the tube forming a curved surface called a "meniscus". Readings are to be taken at the bottom of the meniscus.

All certification exams deal with air conditioning insofar as each specialty exam concentrates on specific aspects of air conditioning. The field of air conditioning can be divided into a number of specialties and each one can be a career choice in itself. Hence, a wide variety of HVAC specialty certifications exist. However, every technician must know a great deal about the general principles of air conditioning. Parallel to a solid knowledge of applied electricity as represented by the vast number of electrical questions, the air conditioning exam is equally important. The following two general air conditioning practice exams of 50 questions each will prepare you for the full-length exam at the end of this guide. A passing score on all exams is 70%.

Many manufacturers, distributors, contractors and government agencies are moving ahead with HVAC certifications. The technician obtaining a general air conditioning certification is ahead of the curve. Technician certification is the HVAC industry's equivalent to the automotive industry's ASC certification for automobile technicians. Technology has brought our industry to a new level of complexity, and qualified technicians are in demand. Technician certification is one way to show you have achieved the level of knowledge and skill required.

AIR CONDITIONING CERTIFICATION PRACTICE EXAM #1

1. Low airflow over a residential split system air conditioning evaporator can cause:

   a. Liquid floodback to the compressor

   b. A decrease in compression ratio

   c. An increase in suction line superheat

   d. Oil logging in the evaporator

2. Which of the following is the correct sequence of work?

   a. Attach the gauge manifold, recover refrigerant, replace the filter core, evacuate the system, recharge the system

   b. Recover the refrigerant, evacuate the system, replace the filter core, recharge the system

   c. Attach the gauge manifold, evacuate the system, replace the filter core, recover the refrigerant, recharge the system

   d. Attach the gauge manifold, recharge the system, evacuate the system, replace the filter core

3. As the pressure on a liquid is increased, the vaporization temperature of the liquid:

   a. Decreases

   b. Increases

   c. Remains unchanged

   d. Varies

4. An oil separator must be mounted:

   a. Level

   b. Below the compressor crankcase

   c. Above the compressor crankcase

   d. In the suction line

Troubleshooting Tech Tip

When making a hole in a duct to insert a thermometer, pitot tube or other instrument, always drill the hole. Never punch the hole! Punching a hole dents the duct and makes the hole more difficult to seal properly.

Instrument Tech Tip

An inclined manometer is more accurate than a digital manometer for measuring velocity pressures. The longer the inclined manometer, the greater the accuracy.

5. Which of the following metering devices controls evaporator pressure?

   a. Thermostatic expansion valve

   b. Fixed orifice

   c. Low side float

   d. None of these

6. The purpose of the external equalizer on a thermostatic expansion valve is:

   a. To equalize evaporator pressure drop with respect to the TEV

   b. To equalize the evaporator pressure drop

   c. To help prevent liquid floodback

   d. To allow the high side pressure to equalize to the low side when the compressor cycles off

7. The higher the MERV rating of a filter-drier, the:

   a. More efficient the drier at adsorbing moisture

   b. Less efficient the drier at adsorbing moisture

   c. Smaller the particles the filter can entrap

   d. MERV does not apply to filter-driers

8. Zeolite is used with:

   a. Refrigerant filter-driers

   b. HEPA filters

   c. Pipe threads

   d. Solder

9. Vacuum pressures are measured in:

   a. Microns

   b. Millimeters of mercury

   c. Inches of mercury

   d. All of the above

10. The number 25,400 relates to:

    a. Microns/inch

    b. Btu/ton

    c. Cfm/hp

    d. Cfm/ton

11. The general rule of thumb for airflow is:

    a. 400 cfm/ton

    b. 400 cfm/hp

    c. 6 cfm/sq'

    d. 144 cfm/cu'

12. A double suction riser is used to:

    a. Help improve refrigerant return

    b. Allow the compressor to pump at a greater capacity

    c. Decrease suction line resistance

    d. Improve oil return as system capacity changes

13. A service technician measures a return air temperature of 80 degrees and a supply air temperature of 55 degrees. What, if anything, may be wrong with this air residential split system air conditioning unit?

    a. Low on refrigerant charge

    b. Low on evaporator airflow

    c. Refrigerant overcharge

    d. There is probably nothing wrong with the system

14. The normal compression ratio for a typical comfort air conditioning application should be in the range of:

    a. 1 to 2

    b. 1.5 to 2.25

    c. 2 to 3

    d. 2.75 to 3.75

Installation Tech Tip

Heating and cooling load calculations are necessary not only to choose the right tonnage of system but also to choose the correct duct runs to deliver the correct amounts of air quantity to be delivered to each room. The total heating and cooling loads are used to select the equipment size and a room-by-room calculation is used to properly size the duct system.

Superheat Tech Tip

Don't confuse thermostatic expansion valve superheat with total low-side system superheat. Thermostatic expansion valve superheat is the difference between the evaporator saturation temperature and the temperature at the thermostatic expansion valve's sensing bulb. Total low-side system superheat is the temperature difference between the evaporator saturation temperature and the temperature of the suction line about 6 inches from the compressor's suction service valve.

15. All else being normal, if a system shows a high low-side superheat with high condenser subcooling, what is most likely the problem?

    a. Refrigerant overcharge

    b. Refrigerant undercharge

    c. A partial refrigerant restriction

    d. There is probably no problem

16. All else being normal, if a system shows a low low-side superheat with high condenser subcooling, what is most likely the problem?

    a. Refrigerant overcharge

    b. Refrigerant undercharge

    c. A partial restriction

    d. There is probably no problem

17. All else being normal, if a system shows a high low-side superheat with low condenser subcooling, what is most likely the problem?

    a. Refrigerant overcharge

    b. Refrigerant undercharge

    c. A partial restriction

    d. There is probably no problem

18. A grille equipped with a damper control is called:

    a. A vane

    b. A register

    c. A louver

    d. A damper

19. A pressure relief valve is:

    a. Direct pressure acting

    b. Remote pressure activated

    c. Manually operated

    d. Reverse pressure acting

20. A parallel compressor system in which both the crankcase oil level and the refrigerant gas pressure are equalized is a:

    a. Double pipe crankcase equalizer system

    b. Single pipe crankcase equalizer system

    c. External equalizer oil system

    d. Crankcase hot gas oil equalizer

21. In a direct expansion chiller used on air conditioning applications, the refrigerant is on what side of the tubes?

    a. Around the diverter

    b. In the tubes

    c. Surrounding the tubes

    d. On both sides

22. The total minimum open area for air to flow through on a supply air register is called:

    a. Free area

    b. Open area

    c. Core area

    d. Drop area

23. Which type of copper tubing has the thickest wall?

    a. Type L

    b. Type K

    c. Type M

    d. Type DWV

24. What is the gross weight of a refrigerant cylinder?

    a. The weight of the cylinder minus the refrigerant

    b. The weight of the cylinder plus the weight of the refrigerant

    c. The weight of cylinder's refrigerant capacity

    d. The weight of the cylinder and the box around the cylinder

Safety Tech Tip

R-22 gauges, manifolds, hoses and recovery equipment should not be used on R-410A systems due to the higher pressure of R-410A. Always use tools and materials specifically rated for use with R-410A on R-410A systems.

Fan Tech Tip

Centrifugal fans that have a single wheel are called single width fans. Single width fans draw air from one side only so they are called single width, single inlet or SWSI fans. Centrifugal fans that have two wheels mounted next to each other on the same motor shaft are called double width fans. They draw air from both ends so they are called double width, double inlet or DWDI fans.

25. What does an anemometer measure?

    a. Feet

    b. Feet per minute

    c. Cubic feet per minute

    d. Vacuum pressure in microns

26. The term "induced draft" most likely refers to:

    a. A compressor

    b. A type of control

    c. A type of cooling tower

    d. A type of test instrument

27. What fitting should be used to connect 1/4" copper tubing to a 1/4" internal pipe thread opening on a compressor crankcase?

    a. A union

    b. A half union

    c. A tee

    d. A street ell

28. Which of the following compressors has the fewest moving parts?

    a. Rotating vane rotary

    b. Fixed vane rotary

    c. Scroll

    d. Reciprocating

29. A hot liquid line with a low to normal high side pressure is an indication of:

    a. An overcharged system

    b. An undercharged system

    c. Air in the system

    d. A restricted filter-drier

30. Air and water are directed counterflow so as to:

    a. Average the flow rate

    b. Gain the greatest amount of heat transfer

    c. Reach the lowest apparatus dew point

    d. Control the humidity

31. Which of the following applications is not a mechanical cooling system?

    a. Cascade system

    b. Reciprocating compression system

    c. Evaporative cooling

    d. Centrifugal compression system

32. A compressor is short cycling. Which of the following would not be the cause?

    a. Low pressure controller differential set too close

    b. Automatic reset high pressure control differential set too close

    c. Low refrigerant charge

    d. Leaking compressor suction valve

33. The position of the valve stem on the suction service valve is normally:

    a. Fully-backseated

    b. Full-frontseated

    c. Midseated

    d. Half cocked

34. Of the three fundamental fan laws, one states that the power varies by the _____ of the speed.

    a. square

    b. cube

    c. rate

    d. inverse

Refrigerant Tech Tip

Systems using R-410A require higher pressure control settings. The high-pressure control opens at 610 psig plus or minus 10 psig and closes at 500 psig plus or minus 15 psig. The low-pressure control opens at 50 psig.

Airflow Tech Tip

VAV terminal units are either "pressure dependent" or "pressure independent." Pressure dependent boxes depend upon the static pressure in the duct for control. Pressure dependent boxes provide much better airflow control because they contain a flow-sensing device that modulates the airflow independent of the duct static pressure.

35. Abnormally low suction pressure can be caused by which of the following?

    a. Dirty air filter

    b. Partially restricted filter-drier

    c. Shortage of refrigerant

    d. Any or all of these

36. Which of the following compressor oil types is generally considered the most compatible with HFC refrigerants?

    a. POE

    b. AB

    c. Mineral

    d. Vegetable

37. The factor determining the correct quantity of air for each room in a building is:

    a. The calculated heat load for each room

    b. The size of the room

    c. The size of the ducts

    d. The customer's preference

38. Refrigerant 22 is classified as an:

    a. A-1 refrigerant

    b. A-2 refrigerant

    c. B-1 refrigerant

    d. B-2 refrigerant

39. Which of the following combinations of instruments can be used to measure airflow in cfm on an electric furnace?

    a. A thermometer, ammeter and voltmeter

    b. A thermometer, tachometer and voltmeter

    c. A tachometer, pitot tube and ammeter

    d. A gauge manifold set, thermometer and voltmeter

40. A rotating vane anemometer measures:

    a. Cfm

    b. Rpm

    c. Fpm

    d. Feet

41. A lockout relay is also called:

    a. An impedance relay

    b. Reset relay

    c. Neither a or b

    d. Both a and b

42. A lockout relay can be reset by:

    a. De-energizing the main power

    b. Pushing the reset button

    c. Turning the breaker off and back on

    d. All of the above

43. A voltage measured across a switch means:

    a. The switch is closed

    b. The switch is open

    c. The switch is bad

    d. The switch is wearing out

44. When installing add-on split system air conditioning for a residence, you should never:

    a. Drill holes through outside cement walls

    b. Fish new thermostat wires through walls where power wires are also located

    c. Fish thermostat wires inside EMT along with power wiring

    d. Install the new evaporator coil in a downflow style furnace cabinet

Compressor Tech Tip

Three-phase scroll compressors must rotate in the correct direction. A reversed three-phase scroll compressor will be noisy, have incorrect operating pressures and may be damaged. The direction of rotation can be changed by reversing any of the three power leads to the compressor.

Installation Tech Tip

Poorly or improperly installed high efficiency heating and cooling equipment will operate at much lower efficiencies than their rating. Always follow the manufacturer's installation, service and operation instructions carefully.

45. In a series circuit, the current:

    a. Is the same throughout the series path

    b. Varies according to the wire size

    c. Is always one-half the voltage

    d. Passes the voltage on its way back

46. Jumping a wire across terminals R and G on the subbase of a thermostat will cause:

    a. The blower motor to operate

    b. The condenser fan to come on

    c. The gas valve to become energized

    d. The compressor to start

47. A fully hermetic compressor on a 3-ton residential split system has three electrical terminals on the side of the compressor. An ohmmeter reads the following: terminal A to B = 20 ohms, A to C = 15 ohms and B to C = 5 ohms. There was a reading of infinity between all three terminals to ground. Which terminal is the common terminal?

    a. Terminal A is common

    b. Terminal B is common

    c. Terminal C is common

    d. There is no common terminal on this type compressor

48. A 3-ton residential split system should move about how many CFM total?

    a. 800 CFM

    b. 1000 CFM

    c. 1200 CFM

    d. 1500 CFM

49. An R-22 air conditioning system is operating with a head pressure of 220 psig and a suction pressure of 68.5 psig. A temperature measurement taken on the suction line 8 inches from the suction service valve reads 60 degrees. A temperature measurement taken on the liquid line leaving the condenser reads 90 degrees F. What is the low side superheat?

    a. 10 degrees of superheat

    b. 15 degrees of superheat

    c. 20 degrees of superheat

    d. 30 degrees of superheat

50. A hot wood stove gives up heat to the room by:

    a. Conduction

    b. Convection

    c. Radiation

    d. All three at the same time

Superheat Tech Tip

Thermostatic expansion valves do not control evaporator temperature or pressure. They control the superheat between the refrigerant saturation temperature and the location of the sensing bulb. Never adjust the valve without taking a superheat measurement.

IAQ Tech Tip

A missing or improperly sized or installed condensate trap can allow water to be blown down the duct, causing mold, moisture and indoor air quality problems as well as allowing insects and unfiltered air to enter the system. Install traps where they are missing and repair those that are incorrectly applied. Inform your customer of the importance of condensate traps.

AIR CONDITIONING CERTIFICATION PRACTICE EXAM #2

1. The Fahrenheit scale is based on boiling water at sea level at what temperature?

    a. 459 degrees

    b. 212 degrees

    c. 180 degrees

    d. 100 degrees

2. Zero pounds gauge corresponds on the absolute scale to:

    a. 144

    b. 212

    c. 14.7

    d. zero

3. Hidden heat in refrigeration work is referred to as:

    a. Intensity of heat

    b. Latent heat

    c. Heat a thermometer can "sense"

    d. Cold

4. Absolute zero on the Fahrenheit scale is:

    a. -459 degrees

    b. -273 degrees

    c. -100 degrees

    d.   0 degrees

5. A ton of refrigeration is a unit equal to:

   a. 2,880,000 btu per day

   b. 12,000 btu per hour

   c. 2,000 btu per minute

   d. All of these are correct

6. The heat used to change a liquid to a gas is called the latent heat of:

   a. Absorption

   b. Vaporization

   c. Fusion

   d. Liquid

7. A thermometer is said to "sense" what?

   a. Heat of fusion

   b. Latent heat

   c. Sensible heat

   d. Specific heat

8. Five pounds of water is heated by two degrees F. How many btu were added to the water?

   a. 25 btu

   b. 10 btu

   c. 5 btu

   d. 2 btu

9. The amount of heat required to melt one pound of ice at 32 degrees F is:

   a. 212 btu

   b. 180 btu

   c. 144 btu

   d. 970 btu

Troubleshooting Tech Tip

A confusing troubleshooting problem may be easier to solve if you make a simple but accurate sketch of the system, find a suitable location to concentrate, and then go over the symptoms and their possible causes as applied to that particular system, its components and its sequence of operation.

Compressor Tech Tip

A compressor should never be allowed to operate with a discharge line temperature exceeding 225 degrees Fahrenheit. Compressor damage and eventual failure will result.

10. Superheat is added:

     a. In changing a liquid to a vapor

     b. In raising the temperature of water

     c. After all the liquid has been changed to vapor

     d. None of these is correct

11. Subcooling is:

     a. Heat added or removed in changing the temperature of a liquid

     b. Heat which causes a liquid to turn into a solid

     c. Heat removed from a liquid below 0 degrees F

     d. Heat removed below the subheat temperature

12. The saturation temperature is:

     a. Never actually reached

     b. When water is at 0 degrees F

     c. The same as the condensing temperature and boiling temperature

     d. The reciprocal of the inverse of the superheat ratio

13. Cold is:

     a. Any temperature below 98.6 degrees F

     b. A temperature lower than 50 degrees F

     c. A relative term with no specific temperature

     d. A temperature near absolute zero

14. As the pressure over a liquid is lowered, _____.

     a. temperature decreases

     b. temperature increases

     c. the boiling point of the liquid increases

     d. the boiling point of the liquid decreases

15. As heat is added to a substance, _____.

    a. the molecules move slower

    b. it becomes even easier to add still more heat

    c. the substance loses heat

    d. the molecules move faster

16. The amount of heat it takes to change the temperature of one pound of a substance by one degree Fahrenheit is called:

    a. Subcooling

    b. Specific heat

    c. Superheat

    d. Specific density

17. The specific heat of a substance changes when:

    a. The temperature is greater than 100 degrees

    b. The substance changes weight

    c. The temperature falls below 30 degrees

    d. The substance changes state

18. A half-filled cylinder of refrigerant at 80 degrees F will be:

    a. Superheated

    b. Subcooled

    c. Saturated

    d. Subheated

19. The air you are breathing is:

    a. Superheated

    b. Subcooled

    c. Saturated

    d. Subheated

IAQ Tech Tip

Sick Building Syndrome (SBS) is a condition considered to be present if at least 20% of the occupants complain of headaches, fatigue, dizziness, nausea, congestion and other related symptoms within a two-week period. If their symptoms go away after leaving the building and return when inside the building, an investigation is warranted.

IAQ Tech Tip

A study of nearly 1000 sick buildings discovered that only 25% had enough outside (fresh) air, only 13% had adequate air filters and only 44% had clean HVAC systems.

20. The three methods of heat transfer are:

    a. Conduction, convection and radiation

    b. Conduction, convection and evaporation

    c. Condensation, convection and radiation

    d. Convection, radiation and sublimation

21. As a gas is compressed, _____.

    a. temperature and pressure increase

    b. pressure decreases and volume increases

    c. temperature decreases and pressure increases

    d. temperature and pressure decrease

22. When compressed enough and then cooled, a gas will:

    a. Explode

    b. Sublimate

    c. Flame off

    d. Condense

23. The condensing temperature and the _____ temperature mean the same thing.

    a. superheated

    b. subcooling

    c. saturation

    d. sublimated

24. The evaporator is a:

    a. High pressure component

    b. Low pressure component

    c. High and low pressure component

    d. None of these is correct

25. The two components that contain coils are:

   a. Condenser and compressor

   b. Condenser and evaporator

   c. Evaporator and compressor

   d. Evaporator and accumulator

26. Refrigerant vaporizes in the:

   a. Condenser

   b. Evaporator

   c. Compressor

   d. Oil separator

27. The two components that divide the high pressure side from the low pressure side of a system are the:

   a. Compressor and condenser

   b. Compressor and metering device

   c. Condenser and evaporator

   d. Liquid line and suction line

28. The pipe which connects the evaporator to the compressor is the:

   a. Suction line

   b. Liquid line

   c. Discharge line

   d. Hot gas line

29. The pipe which connects the compressor to the condenser is the:

   a. Suction line

   b. Liquid line

   c. Discharge line

   d. Hot gas line

Economizer Tech Tip

An economizer not only saves energy during the cooling season, it also improves air quality by bringing in more outside air and allowing older "stale" air to be exhausted from the building. Therefore, servicing and repairing existing economizers is both an economic and indoor air quality issue. Unfortunately, many economizers are inoperable. Repairing them is both a source of revenue and a valuable service to the customer. The cost of the energy saved may cover the repair expenses, and the healthier air quality is of even greater value.

Airflow Tech Tip

The greater the air velocity, the more resistance there is to a change in direction. When a fitting creates turbulence, the turbulence continues for several feet downstream of the fitting.

30. Which one of the following refrigerants would be used with POE oil?

    a. R-22

    b. R-12

    c. R-502

    d. R-410A

31. The discharge line temperature of a reciprocating compressor should never be allowed to exceed how many degrees F?

    a. 190 degrees

    b. 225 degrees

    c. 325 degrees

    d. 375 degrees

32. Refrigerant becomes superheated in the:

    a. Evaporator

    b. Evaporator and suction line

    c. Evaporator, suction line and compressor

    d. Bottom of the condenser

33. Refrigerant is in its saturated state in the:

    a. Compressor

    b. Liquid line

    c. Condenser

    d. Suction line

34. The temperature-pressure chart works:

    a. Where refrigerant is saturated

    b. Where refrigerant is superheated

    c. Where refrigerant is subcooled

    d. Wherever the manifold gauges are attached

35. Which service valve should never be frontseated while the compressor is operating?

    a. Compressor suction service valve

    b. Compressor discharge service valve

    c. King valve

    d. Schrader valve

36. How many microns are there per inch of mercury?

    a. 762,000 microns/inch

    b. 0 microns/inch

    c. 7000 microns/inch

    d. 25,400 microns/inch

37. A manifold gauge set is properly attached to the suction and discharge valves on a compressor. In order to read the high and low side pressures of the system, what position should the service valves on the compressor be in?

    a. Frontseated

    b. Midseated

    c. Backseated

    d. None of these

38. A manifold gauge set is properly attached to the suction and discharge valves on a compressor. In order to read the high and low side pressures of the system, what position should the gauge manifold valves be in?

    a. Fully clockwise

    b. Fully counterclockwise

    c. Midseated

    d. None of these

Gauge Tech Tip

When encountering gauges that have been permanently installed on a system, always remember that vibration, pulsating pressures and subjection to extreme pressures in the past may have rendered the gauges inaccurate. Always check the system with gauges you know are accurate.

Meter Tech Tip

Use meters that are made to measure the RMS (Root Mean Square) value of voltage and amperage. Non-RMS meters measure the average value of a sine wave and may give inaccurate readings when used where electronic devices such as variable frequency drives are installed. Some electronic devices distort the true readings while RMS meters will give correct readings in spite of the distortion.

39. A 125-pound cylinder of R-22 is stored in a room at 80 degrees Fahrenheit. The pressure in the cylinder should be:

    a. 144 psig

    b. 160 psig

    c. 175 psig

    d. 80 psig

40. Vacuum pressures are measured in:

    a. Inches of mercury

    b. Inches of water

    c. Microns

    d. Both a and c

41. The condenser gives up:

    a. Sensible heat

    b. Latent heat

    c. Specific heat

    d. Sensible and latent heat

42. What is the purpose of a valve installed on the high side of the system and which causes the condenser to flood with liquid refrigerant under certain circumstances?

    a. It controls the head pressure during periods of low outside ambient temperatures.

    b. It controls the head pressure during periods of high outside ambient temperatures

    c. It allows the refrigerant to back up into the condenser when the thermostatic expansion valve throttles back

    d. It allows a larger system to operate without the need for a receiver

43. A thermostatic expansion valve controls:

    a. Evaporator pressure

    b. Evaporator temperature

    c. Evaporator superheat

    d. All of the above

44. An automatic expansion valve directly controls:

    a. Evaporator pressure

    b. Evaporator temperature

    c. Evaporator superheat

    d. All of the above

45. The oil failure control is operated by:

    a. The oil pump discharge pressure and the suction pressure

    b. The oil pump discharge pressure and the head pressure

    c. The suction pressure and the head pressure

    d. The oil supply pressure and the oil return pressure

46. A system is running constantly and not cooling the space. The suction line at the compressor is cold and sweating. These symptoms seem to indicate that:

    a. The filter-drier is partially restricted

    b. The compressor is failing

    c. The air filter is restricted and/or the evaporator is dirty

    d. The system has a refrigerant leak

47. A frosted suction line on an air conditioning application indicates:

    a. The system is overcharged

    b. The system is undercharged

    c. The suction line temperature is below the dew point of the ambient air and at or below 32 degrees Fahrenheit

    d. There is air in the system

Meter Tech Tip

Analog meter accuracy is a percentage of the meter's scale. Therefore, the smaller the range of the scale, the greater the accuracy of the meter. Analog meters are handy when measuring constantly changing values. Instead of flashing changing numbers as a digital meter would, the analog needle gently floats over the scale.

48. An oversized air conditioning system may cause:

    a. The system to short-cycle

    b. The humidity to decrease

    c. The compressor to fail before its time

    d. Both a and b

49. Flooded starts on a compressor can be the result of:

    a. Refrigerant migration

    b. Refrigerant overcharge

    c. A burned out crankcase heater

    d. All of the above

50. An air conditioning system has adequate airflow over both the evaporator and condenser. The low side has a superheat of 60 degrees and the condenser subcooling is 25 degrees. What is most likely wrong with this system?

    a. It is overcharged

    b. It is undercharged

    c. It has a restriction in the liquid line

    d. Air is in the system

PART SEVEN
Heat Pump Certification and Practice Exam

What You Will Learn

- What to expect on the heat pump portion of the certification exams

- Confidence and experience as you take and review the 50-question practice heat pump exam

- The confidence and knowledge necessary to deal effectively with the other specialty exams

Safety Tech Tip

MSDS stands for Material Data Safety Sheet. Every hazardous material has such a sheet available which provides information about that material or chemical. MSDS sheets identify the material, the manufacturer, chemical name, ingredients, physical data, fire and/or explosion hazards, health hazards and first aid procedures. Become familiar with the substances you work with and around. Know how to handle the substance properly and what to do in an emergency.

Both NATE and HVAC Excellence offer heat pump certification exams, each consisting of 100 multiple-choice questions. Those passing NATE's heat pump certification exam automatically receive the NATE air conditioning certification at the same time. A technician who is capable of installing, troubleshooting and repairing a heat pump is more than qualified to perform the same services on a cooling-only air conditioning system.

Remember, the NATE certifications expire five years after they are obtained. HVAC Excellence certifications do not expire. The following 50-question practice heat pump exam will give you a good idea of what to expect on a heat pump certification exam. Even the specialty exams, such as this one, may contain some non-heat pump specific questions. Occasionally, entire questions from one of the other specialty exams will appear again on another specialty exam.

HEAT PUMP CERTIFICATION PRACTICE EXAM

1. Charging a heat pump with refrigerant is best accomplished in the _____ mode.

    a. Heating mode

    b. Cooling mode

    c. Defrost mode

    d. Emergency heating mode

2. Heat pumps can provide both heating in the winter and cooling in the summer. In which mode is the heat pump more efficient?

    a. Heating mode

    b. Cooling mode

    c. Defrost mode

    d. Emergency heating mode

3. Which of the following components is exclusive to the heat pump?

    a. Condenser

    b. Oil separator

    c. Reversing valve

    d. Accumulator

4. Which of the following airflow rates is the most reasonable for a heat pump system operating in the heating cycle?

    a. 200 cfm per ton

    b. 300 cfm per ton

    c. 400 cfm per ton

    d. 450 cfm per ton

Safety Tech Tip

Dry skin may have an electrical resistance of over one million ohms. Wet skin may cause skin resistance to drop to less than 300 ohms. Perspiration on a warm day may significantly lower the body's resistance and raise the possibility of a severe electrical shock, burn or even death. An electrical current of only 50 milliamps can be severe while only 100 milliamps can kill.

Safety Tech Tip

To prevent the base of an extension ladder from sliding out from under you when the ground is wet or covered with ice or snow, block the base of the ladder with one of the tires of your service vehicle. If that is not possible, be sure to block the base with any heavy item such as a toolbox.

5. During the heat pump's cooling cycle, the outdoor coil is utilized as a _____ while the indoor coil is utilized as a _____.

   a. Condenser, evaporator

   b. Evaporator, condenser

   c. Compressor, accumulator

   d. Accumulator, condenser

6. Heat is present in winter outdoor air until the outside air ambient temperature reaches a low temperature of:

   a.   0 degrees F

   b.   −32 degrees F

   c. −273 degrees F

   d. −459 degrees F

7. The heat pump utilizes:

   a. Two filter-driers with check valves

   b. One bi-flow filter-drier

   c. A single mono-flow filter-drier in the liquid line

   d. Some use the bi-flow filter-drier and others use two filter-driers with check valves

8. The correct way to size a heat pump is to perform both heating and cooling load calculations and select the heat pump size by:

   a. Adding the heating load and cooling loads

   b. Subtracting the smaller load from the larger load

   c. Using the heating load to size the system

   d. Using the cooling load to size the system

9. Auxiliary electric heat is necessary because:

   a. It provides heat for the home during defrosts

   b. It provides heat for the home in the event of a heat pump failure

   c. It provides heat for the home if the heat pump is unable to meet the demand for heat

   d. All of the above

10. Reading a voltage across the contacts of the outdoor thermostat means:

    a. The thermostat contacts are closed

    b. The thermostat contacts are open

    c. The thermostat contacts are stuck

    d. The thermostat contacts are working correctly

11. Which of the following is not a type of heat pump?

    a. Ground to air

    b. Air to air

    c. Water to air

    d. Direct coupled

12. The coefficient of performance is determined by:

    a. Dividing the output btuh by the input btuh

    b. Dividing the input btuh by the output btuh

    c. Multiplying the input by the output

    d. Subtracting the output from the input

Safety Tech Tip

POE oils are a solvent and can be harmful to the skin and eyes. They can also damage some roofing materials. Do not come into contact with the oil and be careful to avoid spills.

Sheet Metal Tech Tip

Avoid choked sheet metal fittings on installations as they have high resistance to flow and create additional flow turbulence. A fitting is chocked if the area in the middle of the fitting is less than the area of its ends.

13. The "balance point" of a heat pump is:

    a. The lowest temperature at which the heat pump can meet the heating requirement for the structure

    b. The lowest temperature at which the heat pump can operate

    c. The temperature at which the heat pump has the highest efficiency

    d. The point at which the heating and cooling loads meet

14. Which of the following is an incorrect term for the solenoid valve that switches the heat pump between the heating and cooling modes?

    a. Change-over valve

    b. Reversing valve

    c. Four-way valve

    d. Discharge operator valve

15. The best location for the outdoor section of an air to air heat pump is:

    a. Where it gets the winter sun

    b. Where it is sheltered from the sun's radiation

    c. Where it never snows

    d. Under the shelter of a porch

16. What may happen if the body of a reversing valve is slightly dented?

    a. Nothing, as they are made to be very durable

    b. The internal slide may stick

    c. The valve may make noises from the gas trying to pass through partially open ports

    d. Both b and c are correct

17. The reversing valve has 4 pipe connections. One pipe connection is on one side of the valve and three pipe connections are on the opposite side. What is the center connection on the side with three pipe connections?

    a. Compressor suction connection

    b. Compressor discharge connection

    c. Outdoor coil connection

    d. Indoor coil connection

18. The reversing valve has 4 pipe connections. One pipe connection is on one side of the valve and three pipe connections are on the opposite side. What is the single connection on the side opposite the three connections?

    a. Compressor suction connection

    b. Compressor discharge connection

    c. Outdoor coil connection

    d. Indoor coil connection

19. Why are reversing valves pilot operated?

    a. Because pilot operated valves use the differential pressure to operate the valve ports instead of having a solenoid valve working against the head pressure

    b. Because that is just the way they are made

    c. Pilot operated valves are quiet

    d. Heat pump reversing valves are NOT pilot operated!

Customer Tech Tip

We have all seen it many times – a service vehicle with paperwork, trash, parts and materials piled up on the dashboard. The message it sends is that the owner or user is messy and disorganized. Present a better image to your customers by keeping your dashboard clear of everything. Do not use it as a shelf.

Maintenance Tech Tip

When stopping a system

which employs a VFD (Variable

Frequency Drive) always

de-energize the VFD before

turning off the system power

to prevent possible damage to

the VFD.

20. Two technicians are working together on a heat pump system and are preparing to evacuate the system. Technician A says that in order to pull a good vacuum the reversing valve needs to be switched on and off during the evacuation so air and moisture are not trapped in the system. Technician B says technician A is incorrect and that the reversing valve will not switch with the system in a vacuum anyway. Who is correct?

    a. Technician A is correct

    b. Technician B is correct

    c. Both technicians A and B are correct

    d. Both technicians A and B are incorrect

21. Terminal "O" on a heat pump thermostat is used for:

    a. Operating the reversing valve

    b. Operating the indoor fan motor

    c. Operating the compressor

    d. Operating the condenser fan motor

22. Terminal "G" on a heat pump thermostat is used for:

    a. Operating the reversing valve

    b. Operating the indoor fan motor

    c. Operating the compressor

    d. Operating the condenser fan motor

23. A typical 3-ton split system residential heat pump would have auxiliary electric heating elements of about how many watts total?

    a. 4 kw

    b. 5 kw

    c. 15 to 20 kw

    d. 40 to 60 kw

24. Typical slide-in packages of electric heating elements come in units of how many kilowatts each?

    a. 1 to 2 kw

    b. 3 to 6 kw

    c. 15 kw

    d. 25 kw

25. When a typical heat pump goes into defrost which of the following will occur?

    a. The compressor shuts off, the reversing valve switches to cooling, the defrost heaters are energized and the outdoor fan stops

    b. The compressor remains on, the outdoor fan stops and the auxiliary electric elements are energized

    c. The indoor and outdoor fans operate at a higher speed, the compressor remains on and the auxiliary electric elements remain off

    d. The auxiliary electric elements operate at a reduced voltage, the compressor runs at half speed and the outdoor fan shuts off

26. What function does a Ranco E-15 control serve?

    a. It is a common, combination time and temperature defrost control

    b. It is a mechanical defrost timer with a built-in thermostatic control

    c. It is a defrost control that checks the outdoor coil for frost build-up every 30, 45 or 90 minutes

    d. All of the above are correct

Customer Tech Tip

When a customer is especially grateful for your service and tells you so, ask the customer to refer your company to his or her friends and relatives. Personal referrals are the best advertising, outweighing any marketing campaign that money can buy.

Safety Tech Tip

Negative interior building pressure with respect to the outside of a structure can be very dangerous to occupants. Negative building pressure can cause fuel burning appliances and heating systems to burn inefficiently due to a lack of oxygen available for complete combustion. Furthermore, the negative pressure may prevent the proper venting of those same products of combustion and distribute them throughout the building. Dangerous products of combustion include carbon monoxide, nitric oxide and nitrogen dioxide.

27. Heat pumps often employ which of the following accessories?

   a. An accumulator
   b. Evaporator pressure regulator
   c. Electronic suction throttling valve
   d. Crankcase pressure regulator

28. Compressor crankcase heaters are used to:

   a. Help decrease refrigerant migration
   b. Reduce the potential of liquid floodback
   c. Reduce the potential of flooded starts
   d. Both a and c are correct

29. A heat pump is operating in the heating mode. The lower the outside air temperature, the:

   a. More efficient the heat pump becomes
   b. Higher the SEER rating
   c. Higher the compression ratio
   d. Lower the balance point

30. Which of the following can prevent the reversing valve from changing positions?

   a. Low refrigerant charge
   b. Overcharge of refrigerant
   c. Air in the system
   d. Low indoor airflow due to a dirty air filter

31. The supply air temperature on a normally operating heat pump is:

    a. Lower than that of a gas heating system

    b. Higher than that of a gas heating system

    c. About the same as that of a gas heating system

    d. Heat pumps do not use supply air

32. When a heat pump is operating in the cooling mode, the suction line at the compressor should feel _____ and when operating in the heating mode the same suction line should feel _____.

    a. Cool, hot

    b. Hot, cool

    c. Hot, hot

    d. Cool, cool

33. High SEER rated heat pumps are those that employ:

    a. Scroll compressors and oversized coils

    b. Reciprocating compressors and oversized coils

    c. Scroll compressors and accumulators

    d. Reciprocating compressors and accumulators

34. Which of the following metering devices can improve the operating efficiency of a heat pump?

    a. Thermostatic expansion valve

    b. Automatic expansion valve

    c. Fixed orifice

    d. Capillary tube

Certifcation Tech Tip

Become a certified technician in each of the areas in which you work. Encourage others to do so as well. Continue to take recognized continuing education classes and seminars to stay current with changes in the industry. New technologies, products, equipment and tools are constantly being developed. Share your new knowledge and skills with others and learn from others as well.

35. During the defrost cycle the outdoor fan motor is:

    a. Shut down

    b. Operated at a reduced speed

    c. Stepped up to its highest speed

    d. Allowed to run at its normal speed

36. Which of the following compressors is most "liquid slug" tolerable?

    a. Reciprocating

    b. Scroll

    c. Compliant scroll

    d. Rotary

37. In the heating mode, ice can still collect on the outdoor coil even when the outside air temperature is above 40 degrees. Why?

    a. Because the outside air temperature of 40 degrees still has a dew point temperature below the 32 degree freezing point

    b. Because the outdoor coil is vaporizing refrigerant at a temperature below 32 degrees

    c. Because the latent heat of the refrigerant is always below freezing

    d. The statement that the coil will continue to ice even though the outside air temperature is above 40 degrees is an incorrect statement

38. The outdoor coil on a heat pump is defrosted using:

    a. Hot gas

    b. Electric heaters

    c. Off cycle defrost

    d. Straight time defrost

39. A customer is concerned that during the heating season heavy ice builds up on one side of the concrete equipment pad and coil. The rest of the coil is clear of ice and the heat pump is meeting the heating needs of the house. What is most likely the cause of the problem?

    a. High winds on that side of the outdoor unit

    b. The outdoor unit is not level

    c. There is more refrigerant on that side of the coil

    d. A defrost problem exists

40. SEER is an abbreviation for:

    a. Seasonal Energy Excellence Rating

    b. Seasonal Energy Efficiency Ratio

    c. Summer Energy Efficiency Ratio

    d. Special Efficiency Energy Rating

41. Typical residential split system heat pumps operate on how many voltages?

    a. One

    b. Two

    c. Three

    d. Four

42. The United States Department of Energy has mandated that the minimum allowable SEER rating for a new residential air conditioning system installation is:

    a. 11 SEER

    b. 12 SEER

    c. 13 SEER

    d. 14 SEER

Safety Tech Tip

Do not take chances with compressed gases. Nitrogen and oxygen cylinders are typically under pressures as high as 2000 psig and require the use of regulators. Acetylene pressures are as high as 250 psig. Never allow oxygen to come into contact with oil or any other similar substances and never use oxygen for pressure testing or purging.

Study Tech Tip

Correct spelling and grammar on estimates and service work orders are more important than many technicians understand. Customers judge and determine their level of trust in the competence and ability of their technician by how professional the technician comes across. Correct spelling and grammar are a key component of professionalism.

43. Which of the following actions will result in a decrease in volumetric efficiency?

   a. An increase in head pressure

   b. A decrease in head pressure

   c. An increase in suction pressure

   d. Both a and c

44. A heat pump operating in the cooling mode would be expected to have a return-to-supply air temperature differential in the range of:

   a. 5 to 6 degrees

   b. 5 to 8 degrees

   c. 8 to 12 degrees

   d. 18 to 22 degrees

45. A heat pump operating normally in the heating mode would be expected to have a supply air temperature in the range of:

   a. 70 to 80 degrees

   b. 80 to 90 degrees

   c. 85 to 105 degrees

   d. 100 to 130 degrees

46. In any case, the compressor's discharge line temperature should not exceed what temperature?

   a. 175 degrees F

   b. 200 degrees F

   c. 225 degrees F

   d. 350 degrees F

47. A heat pump defrosts even when there is no ice on the outside coil. Which of the following is most likely the problem?

    a. The defrost time clock is set for too short a cycle

    b. The outdoor coil is dirty and the system uses a pressure differential type of defrost initiation

    c. The sensing bulb of the defrost control is mounted too close to the bottom of the outdoor coil

    d. Each of these could cause the problem

48. A technician places a heat pump into the auxiliary electric heating mode with the compressor de-energized. He measures a temperature rise between the return and supply air of 40 degrees. The measured voltage at the electric elements is exactly 222 volts and the measured amperage of the heaters is 60 amps. Using the equation CFM = BTU divided by $1.08 \times$ TD, what is the CFM of the system?

    a. 825 CFM

    b. 950 CFM

    c. 1055 CFM

    d. 1500 CFM

49. Solid state defrost control boards can be adjusted to:

    a. Program the thermostat from outside the house

    b. Change the frequency of defrosts

    c. Control the speed of the outdoor fan during defrost

    d. Control the speed of the indoor fan during defrost

Study Tech Tip

Temperature is a measure of heat intensity and the BTU a measure of heat quantity.

Study Tech Tip

Customers notice neat work as well as the organization of a service vehicle. The professional service technician keeps the dashboard of his or her vehicle clear of tools, parts, parts boxes, invoices and other clutter.

50. A thermistor is an:

    a. Electronic temperature sensor that changes its resistance with changes in temperature

    b. Electronic pressure sensor that changes its resistance with changes in pressure

    c. Electro-mechanical thermostat that is used to terminate the defrost cycle when the outdoor coil is clear of ice and frost

    d. None of these is correct

PART EIGHT
Gas Heating Certification and Practice Exam

What You Will Learn

- What to expect on the gas heating certification exams

- Confidence and experience as you take and review the 50-question practice gas heating certification exam

Combustion Tech Tip

It takes 15 cubic feet of air to properly burn one cubic foot of natural gas. Each cubic foot of air has a 21% oxygen content. Therefore, it requires a great deal of air to provide enough oxygen for complete combustion. A source of sufficient air for combustion is essential.

Under normal circumstances, heating and cooling are seasonal. Few technicians are able to work on air conditioning cooling systems year-round. Most must also install and service heating systems during the heating season. Although technicians in some geographic areas may be able to service air conditioning systems in the summer and heat pumps in the winter, most technicians must be able to service gas heating systems as well. Most technicians will want to become certified in gas heat as well as air conditioning and heat pumps. Of course, technicians who become NATE certified in heat pumps are automatically certified in air conditioning. However, to become certified in gas heat, you must pass the NATE or HVAC Excellence gas heat certification exam. The 50-question, multiple-choice practice exam in this section will give you an excellent idea of what the actual exams are like and will help you prepare to pass the real thing.

Gas heat remains a standard method of heating homes and buildings. The increases in gas combustion efficiencies has made gas heating an even more viable and economic source of heat. In addition, the inherent dangers of working with any fuel require a knowledgeable and safe technician to perform the work. There is a strong need for qualified gas heating technicians able to properly install, troubleshoot and service gas fired systems. Obtaining a gas heat certification is one way to show your qualifications.

GAS HEATING CERTIFICATION PRACTICE EXAM

1. Since 1992, the United States Department of Energy has required that all furnaces sold and installed from that date on must have an AFUE of no less than:

   a. 70%

   b. 75%

   c. 78%

   d. 80%

2. Many pre-1992 furnaces have efficiencies around _____.

   a. 10%

   b. 40%

   c. 60%

   d. 80%

3. Today, furnaces with AFUE ratings greater than _____ are available.

   a. 85%

   b. 90%

   c. 100%

   d. 110%

4. A thermocouple is used with:

   a. Intermittent pilot systems

   b. Direct spark ignition systems

   c. Glow coil ignition systems

   d. Standing pilot ignition systems

Air Quality Tech Tip

A tightly constructed building must have a source of outside air to prevent the house from operating in a negative pressure. Fuel-burning appliances such as ranges and hot water heaters consume air in the house and deplete the oxygen level unless a source of outside air is available. Bathroom and kitchen exhaust fans can make the problem even worse, creating an unhealthy living environment.

Safety Tech Tip

Propane and butane fuel gases are heavier than air, collect near the floor and do not easily dissipate. Therefore, they are easily ignited by a spark, the hot surface of an automobile, a pilot light on a water heater or other ignition source.

5. The purpose of the gas pressure regulator is to:

    a. Keep the color of the flame constant

    b. Keep the high limit switch from tripping

    c. Maintain a constant gas pressure at the burner

    d. Allow the gas valve to operate smoothly and accurately

6. The instrument used to measure outlet gas pressure is the:

    a. Velometer

    b. Anemometer

    c. Monometer

    d. Draft gauge

7. The regulator on a natural gas furnace should be adjusted to supply _____ inches of pressure at the burners.

    a. 11"

    b. 10"

    c. 3.5"

    d. 2"

8. Burning one cubic foot of natural gas releases approximately _____ btu.

    a.   800

    b. 1,050

    c. 12,000

    d. 15,000

9. Under ideal conditions natural gas has an ignition temperature of _____ degrees F and a burning temperature of _____ degrees F.

    a. 800 degrees, 1000 degrees

    b. 1,100 degrees, 3,500 degrees

    c. 1,100 degrees, 5,000 degrees

    d. 2,500 degrees, 8,000 degrees

10. Propane and butane are both:

    a. Lighter than air

    b. Heavier than air

    c. Weigh the same as air

    d. None of the above

11. In order to achieve complete combustion with natural gas, each cubic foot of gas must combine with no less than _____ cubic feet of air with an additional 5 cubic feet of extra air to ensure there is enough oxygen for complete combustion.

    a. 5 cubic feet

    b. 10 cubic feet

    c. 15 cubic feet

    d. 20 cubic feet

12. The three elements of combustion are:

    a. Fuel, oxygen and ignition

    b. Nitrogen, oxygen and heat

    c. Air, fuel and oxygen

    d. Air, fuel and nitrogen

Electrical Tech Tip

The normal voltage output of a thermocouple is 15 to 30 millivolts. Like light bulbs, thermocouples and hot surface ignitors burn out after many hours of use. Like light bulbs, you cannot predict when they will burn out. Replacing them after every two heating seasons is a good practice to follow. While doing this, perform a complete seasonal heating service.

Ignition Tech Tip

Hot surface ignitor operating temperatures range from 1500 to 2000 degrees Fahrenheit. Under ideal conditions, the ignition temperature of natural gas is about 1100 degrees Fahrenheit.

13. Which of the following is a natural result of combustion?

   a. Water vapor

   b. Carbon monoxide

   c. Carbon dioxide

   d. All of the above

14. Primary air is air that:

   a. Is mixed with the fuel before the fuel reaches the burner

   b. Is mixed with the fuel at the burner after the fuel has ignited

   c. Remains unused after the combustion process is complete

   d. Enters the draft diverter and goes up the stack with the products of combustion

15. Secondary air is air that:

   a. Is mixed with the fuel before the fuel reaches the burner

   b. Is mixed with the fuel at the burner after the fuel has ignited

   c. Remains unused after the combustion process is complete

   d. Enters the draft diverter and goes up the stack with the products of combustion

16. The three major furnace configurations are:

   a. Upflow, downflow and lowboy

   b. Upflow, downflow and horizontal

   c. Upflow, downflow and crossflow

   d. None of these is correct

17. Residential gas furnaces use blowers powered by which types of motors?

    a. Polyphase, two phase, single phase and capacitor start

    b. Shaded pole, split-phase, capacitor start and PSC

    c. Three phase, shaded pole and PSC

    d. Split-phase, induction start-induction run and variable speed DC

18. The horsepower range of residential heating blower motors is:

    a. 1⁄8 to 1⁄16 hp

    b. 1⁄4 to 1⁄2 hp

    c. 1⁄2 to 2 hp

    d. 3⁄4 to 2 hp

19. A heat anticipator is wired:

    a. In series with the "W" terminal

    b. In parallel with the "W" terminal

    c. In series with the "R" terminal

    d. In parallel with the "R" terminal

20. Placing a jumper wire between the "R" terminal and the "G" terminal on a thermostat subbase will energize the:

    a. Gas valve

    b. Indoor fan motor

    c. Outdoor fan motor

    d. Compressor

Furnace Tech Tip

All high-efficiency condensing furnaces use induced draft blowers. However, not all furnaces using induced draft blowers are high-efficiency furnaces. The highly efficient condensing furnaces employ a secondary heat exchanger with flue gases less than 275 degrees and a flue gas condensate drain line.

Draft Tech Tip

The typical over-the-fire draft measurement for gas fired heating systems should be approximately .01 to .02 inches WC and the typical stack draft should be about .02 inches WC for proper operation.

21. Placing a jumper wire between the "R" terminal and the "W" terminal on a thermostat subbase will energize the:

    a. Gas valve

    b. Indoor fan motor

    c. Outdoor fan motor

    d. Compressor

22. Carbon monoxide is a:

    a. Product of incomplete combustion

    b. Poisonous gas

    c. Odorless gas

    d. All of the above are correct

23. The chemical symbol for carbon monoxide is:

    a. CO_2

    b. CO

    c. H_2O

    d. NH_3

24. Selecting a lower speed tap setting on a furnace indoor blower motor will cause:

    a. The temperature rise through the furnace to increase

    b. The temperature rise through the furnace to decrease

    c. The temperature rise through the furnace will not be affected

    d. The furnace to act as a humidifier

25. "Carry over" channels or "cross overs" are used to:

   a. Carry unused excess fuel back to the flue where it is sent up the stack

   b. Carry primary air to the heat exchanger

   c. Carry secondary air to the heat exchanger

   d. Carry the flame from the first burner that lights to the remaining burners

26. Modern high-efficiency burners use combustion blowers. What are the two major types of combustion blowers?

   a. Induced draft and forced draft

   b. High pressure and low pressure

   c. Impeller type and centrifugal type

   d. Liquid and vapor types

27. The purpose of the combustion blower is to:

   a. Cause an adequate amount of oxygen-rich air to mix with the fuel for better combustion

   b. Push the products of combustion out of the building

   c. Prevent carbon monoxide from entering the heat exchanger

   d. Give the larger blower motor a rest

28. Heat transfers from an object or area at a higher temperature to an object or area at a lower temperature. A heated heat exchanger transfers heat to the air by:

   a. Conduction

   b. Natural convection

   c. Radiation

   d. Forced convection

Combustion Tech Tip

The four types of air necessary for combustion are primary air, secondary air, excess air and dilution air. Sufficient amounts of all four are necessary for complete combustion and the health and safety of the occupants.

Troubleshooting Tech Tip

Medium and high-efficiency

furnaces will not operate properly

if they are not well grounded or

if the incoming power is not

correctly polarized.

29. The voltage that creates a spark on a spark ignition system is approximately:

   a.   240 volts

   b.   5,000 volts

   c. 15,000 volts

   d. 25,000 volts

30. The capacity of a natural gas furnace is de-rated by _____ for every 1000 feet above sea level:

   a. 1%

   b. 2%

   c. 3%

   d. 4%

31. A crack in a heat exchanger can:

   a. Allow combustion gases to mix with the space air

   b. Prevent the heat from effectively transferring to the air

   c. Cause the high-limit control to shut the furnace down

   d. Cause the furnace to catch on fire

32. Gas valves operate on what voltage?

   a.   12 volts

   b.   24 volts

   c. 120 volts

   d. 230 volts

33. Mid-efficiency furnaces operate with flue gas temperatures between:

    a. 275 and 360 degrees

    b. 200 and 500 degrees

    c. 100 and 275 degrees

    d. 80 and 100 degrees

34. High-efficiency furnaces operate with flue gas temperatures between:

    a. 275 and 360 degrees

    b. 200 and 500 degrees

    c. 100 and 275 degrees

    d. 80 and 100 degrees

35. "B" vent is:

    a. Hard, inflexible white plastic vent tubing able to withstand temperatures up to 325 degrees F

    b. Hard, inflexible white plastic vent tubing able to withstand temperatures up to 400 degrees F

    c. Single wall sheet metal vent pipe

    d. Double wall sheet metal vent pipe

36. Latent heat is:

    a. Heat that causes a change in temperature with no change in state

    b. Heat that causes a change in state with no change in temperature

    c. The large amount of heat involved when a gas condenses into a liquid

    d. Both b and c are correct

Efficiency Tech Tip

All else being equal, a VFD (Variable Frequency Drive) is a more efficient method of controlling airflow than IGV (Inlet Guide Vanes). Even though the initial cost of a VFD is higher, the difference is often quickly recovered by the energy saved.

Flue Tech Tip

Damaged, rusted or poorly attached flue pipe and connections must be replaced. Flue gases are highly dangerous and contain carbon monoxide, the silent killer, which is only detectable using special instrumentation.

37. A mid-efficiency or high-efficiency furnace will not operate if:

    a. It is not grounded or the polarity is reversed

    b. The indoor blower motor capacitor is wired into the wrong motor winding

    c. The air filter is installed backwards

    d. The door switch is shorted closed

38. Furnaces are placed into categories based upon:

    a. Venting characteristics

    b. Btuh capacity range

    c. Gas valve type

    d. Total airflow in CFM

39. Condensing furnaces:

    a. Condense water vapor out of the air at the burner, thus increasing the efficiency of the burn

    b. Increase the combustion efficiency by adding condensed oxygen to the fuel

    c. Squeeze additional heat out of the flue gases by removing the latent heat of condensation from the gas and causing much of the gas to liquefy

    d. None of the above is correct

40. Secondary heat exchangers:

    a. Are optional heat exchangers used when necessary

    b. Are constructed so the return air passes over the secondary heat exchanger first and then passes over the main heat exchanger

    c. Are used to add still more heat to the supply air as the supply air makes its last pass before entering the sheet metal ducting system

    d. Are now rarely used on the higher efficiency furnaces

41. A combination fan-stat and high limit:

    a. Are both safety controls

    b. Consists of a high temperature safety thermostat built into the wall thermostat

    c. Are separate switches operating off a single helical bimetal with one switch controlling the indoor fan and the other acting as a high temperature safety control

    d. All the above are correct

42. The three basic categories of ignition systems are:

    a. Continuous ignition, intermittent ignition and direct ignition

    b. Continuous ignition, direct ignition and manual ignition

    c. Automatic ignition, intermittent ignition and direct ignition

    d. Automatic ignition, direct ignition and indirect ignition

Temperature Tech Tip

Saturation temperature is the temperature at which a liquid turns to a vapor or a vapor turns to a liquid. Vapor above the saturation temperature is superheated. A liquid below the saturation temperature is subcooled.

43. The intermittent ignition system:

   a. Lights the burner directly

   b. Uses a thermocouple

   c. Lights the pilot only on a call for heat

   d. Is the most efficient ignition system

44. Direct ignition:

   a. Uses a spark or hot surface igniter to light the burner directly

   b. Gets its name from the fact that it is immediate, direct or fast acting

   c. Lights the pilot only on a call for ignition

   d. Is the most efficient ignition system

45. The flame sensor:

   a. Is a thick ceramic rod that senses the heat of the flame and causes the rod to expand and signal the solid state flame board that a flame is present

   b. Is a small-diameter conducting rod that senses when a flame is present because a current flow moves between the rod and the burner when the flame creates an electrical path between them

   c. Is actually a metal-oxide diode that rectifies alternating current to direct current for the gas valve

   d. Is a device that shuts down the main gas valve if the burner flame gets out of control and starts burning outside the fire box or combustion chamber

46. Which of the following is a typical sequence of operation for a high-efficiency condensing furnace?

   a. Upon a call for heat the induced draft blower starts and runs for 20 seconds, the ignitor heats for 15 seconds, the gas valve opens, the burner lights off, the flame sensor detects a burner flame and monitors the flame for the entire heating cycle until the thermostat is satisfied and the system shuts down.

   b. Upon a call for heat the ignitor heats up for 15 seconds, the gas valve opens and the induced draft blower starts, the burner lights and then the flame sensor takes over monitoring the burner flame until the thermostat is satisfied and the system shuts down.

   c. Upon a call for heat the induced draft blower starts and runs for 20 seconds, the ignitor heats up for 15 seconds, the gas valve opens, the burner lights off and the flame sensor detects and monitors the burner flame. When the plenum heats up, the blower starts and pulls return air over the heat exchanger. When the thermostat is satisfied the system shuts down and the blower shuts off soon after as the plenum cools.

   d. None of these is correct, as condensing furnaces do not use induced draft blowers.

47. A yellow burner flame is usually caused by:

   a. A weak ignitor

   b. Lack of sufficient air

   c. Low gas pressure

   d. Too much primary air

48. A lifting burner flame is usually caused by:

   a. A new ignitor that has not been broken in

   b. High gas pressure and/or too much primary air

   c. Low gas pressure and/or not enough primary air

   d. Blockage in the flue

Notes

Notes

49. What is the purpose of "timing a gas meter"?

    a. To determine the actual input in Btuh of the furnace

    b. To determine if the gas meter is accurate

    c. To check the output of the gas valve

    d. To test for carbon monoxide

50. Electrical troubleshooting of a furnace is best accomplished with a:

    a. Pictorial wiring diagram

    b. Photographs

    c. Voltmeter

    d. Both a and c

PART NINE
Air Distribution, Measurement and Balancing Certification with Practice Exam

What You Will Learn

- What to expect on the air side exams

- Confidence and experience as you take and review the practice air distribution certification exam

Airflow Tech Tip

The aspect ratio is the larger side of the duct in inches divided by the shorter side of the duct in inches. A 24-inch by 24-inch duct has an aspect ratio of 1:1. A 20-inch by 10-inch duct has an aspect ratio of 2:1. The greater the aspect ratio, the greater the resistance to airflow. Ducts should be constructed to keep the aspect ratio under 3:1. A round duct has the least friction loss, square duct is next best and, as the aspect ratio increases, the friction loss increases accordingly. Ducts with high aspect ratios also use more material.

An air conditioning system is made up of three sub-systems: the mechanical system, electrical system and air system. The mechanical system includes the components, refrigerant and piping of the mechanical refrigeration cycle. The electrical system includes all the power and control components necessary to power and operate the mechanical and air systems. The air system includes the fan, ductwork, sheet metal fittings, grilles, registers and the air with its changing properties. A well-educated technician will have the knowledge and ability to correctly recognize, analyze and troubleshoot the air system using air measurement instruments and basic mathematics. The air system is perhaps the least understood of the three sub-systems.

In addition to the general NATE and HVAC Excellence certifications related to air distribution, there are several more sophisticated air and water balancing certifications for those specializing in those areas. TABB—Testing, Adjusting and Balancing Bureau, and NEBB—National Environmental Balancing Bureau are two different certification organizations, each specializing in certifying balancing technicians. These certifications are beyond the scope of this book. However, the general airside topics that are often tested are covered in this chapter.

AIR DISTRIBUTION CERTIFICATION PRACTICE EXAM

1. The density of air is:

   a. How much air weighs per cubic foot

   b. How much volume one pound of air occupies

   c. The ratio of the weight of water to that of the same volume of water

   d. The ratio of the amount of water vapor contained in a cubic foot of air to the amount of moisture it could hold at saturation

2. The specific volume of air is:

   a. How much air weighs per cubic foot

   b. How much volume one pound of air occupies

   c. The ratio of the weight of water to that of the same volume of water

   d. The ratio of the amount of water vapor contained in a cubic foot of air to the amount of moisture it could hold at saturation

3. The relative humidity of air is:

   a. The percent of moisture air is holding as opposed to the amount it could hold if it were saturated

   b. Always less than 100%

   c. The most accurate measurement of how much water vapor is in a given body of air

   d. Used as a multiplier to size humidifiers

4. As air is heated it:

   a. Contracts

   b. Expands

   c. Does not change its volume

   d. Becomes more dense

Airflow Tech Tip

CFM equals Area in square feet times Velocity in feet per minute. Velocity equals CFM divided by Area. Area equals CFM divided by Velocity.

Airflow Tech Tip

Undersized ducts deliver less air at higher velocities and create air noise. Increasing the fan speed to move more air creates more noise. Properly sized ducts deliver the proper quantity of air at a reasonable velocity, are quieter and require less fan energy.

5. The air pressure in a sealed balloon is:

    a. Static pressure

    b. Velocity pressure

    c. A combination of static and velocity pressures

    d. None of the above

6. A blower connected to a length of duct:

    a. Creates static pressure which is then converted to velocity pressure

    b. Creates velocity pressure directly

    c. Creates resistance to the flow of air

    d. None of the above

7. Velocity pressure is determined by:

    a. Subtracting static pressure from total pressure

    b. Subtracting total pressure from static pressure

    c. Dividing the total pressure by the static pressure

    d. Subtracting the traverse pressure from the velocity

8. Velocity pressure can be mathematically converted to:

    a. CFM

    b. FPM

    c. Relative humidity

    d. Inches of mercury vacuum

9. An inclined manometer is basically:

   a. A U-tube manometer tilted at a steep angle

   b. A hygrometer on a stand

   c. A digital micron gauge

   d. A type of sling psychrometer

10. A duct measures 24 inches by 18 inches. The air velocity is 600 FPM. How many CFM are being delivered through the duct? Round the number to the nearest whole number:

    a. 1600 CFM

    b. 1800 CFM

    c. 2000 CFM

    d. 2200 CFM

11. The static pressure of a duct is 1.76 inches and the total pressure is 2 inches. What is the velocity pressure?

    a. .024 inches

    b. .24 inches

    c. 3.76 inches

    d. zero

12. A pitot tube can directly measure:

    a. Total and velocity pressures

    b. Total and static pressures

    c. Velocity and static pressures

    d. None of the above

Airflow Tech Tip

Single wall turning vanes properly installed in a sheet metal elbow reduce the dynamic loss and improve airflow through the fitting. Poorly installed turning vanes create more resistance than no vanes.

Airflow Tech Tip

If duct system resistances and the resulting pressure losses are kept low, less energy will be needed to operate the blower and operating costs will be less.

13. Which of the following is the best description of a duct traverse?

    a. A series of carefully spaced velocity pressure readings which may be converted to fpm using an equation

    b. A series of carefully spaced velocity pressure readings which may be converted directly to cfm using an equation

    c. A series of carefully spaced static pressure readings which may be converted to fpm using an equation

    d. A series of carefully spaced static pressure readings which may be converted to cfm using an equation

14. How many holes must be made in a round duct in order to take a series of traverse readings?

    a. One

    b. Two

    c. Three

    d. Four

15. A thermo-anemometer measures:

    a. CFM directly

    b. FPM directly

    c. Total and static pressures directly

    d. Temperature and humidity

16. The dew point temperature is:

    a. The temperature at which water vapor in the air will condense into a liquid

    b. The temperature at which air gives up its moisture

    c. The saturation temperature of water vapor in a body of air

    d. All of the above

17. Dehumidification occurs when an air conditioning coil:

    a. Operates at or below the dew point of the air moving across it

    b. Operates above dew point

    c. The cooling coil always removes moisture as long as it is operating

    d. Cooling coils do not dehumidify air

18. Who usually determines if a duct leakage test is to be performed and the test pressure at which it will be performed?

    a. The design engineer

    b. The installing contractor

    c. The TAB technician

    d. The sheet metal foreman

19. When testing a duct system to determine its leakage rate, why is a special test stand necessary?

    a. It contains a calibrated orifice and matching flow chart for that orifice

    b. It is manufactured to accepted testing standards

    c. Test results can be traced to the specific test stand used for the test if there are concerns about accuracy

    d. All of the above

20. Why are two manometers mounted on a duct leakage test stand?

    a. To use one as a spare in case something happens to the other

    b. To be checked for accuracy against each other

    c. To speed up the testing procedure

    d. None of these is correct

Airflow Tech Tip

Flexible duct is comparatively rough on the inside, adding a great deal of resistance and pressure loss to the run. Most building codes limit the length of run for flexible duct.

Airflow Tech Tip

The inclined manometer is used for very accurate air pressure readings and for calibrating other air measurement instruments.

21. When the dry bulb temperature increases, the relativity humidity of air:

    a. Increases

    b. Decreases

    c. Remains the same

    d. None of these

22. When the dry bulb temperature increases, the wet bulb temperature:

    a. Increases

    b. Decreases

    c. Remains the same

    d. None of these

23. When the dry bulb temperature increases, the dew point temperature:

    a. Increases

    b. Decreases

    c. Remains the same

    d. None of these

24. When the dry bulb temperature increases, the specific humidity:

    a. Increases

    b. Decreases

    c. Remains the same

    d. None of these

25. When the dry bulb temperature increases, the specific volume of the air:

    a. Increases

    b. Decreases

    c. Remains the same

    d. None of these

26. When the dry bulb temperature increases, the density of air:

    a. Increases

    b. Decreases

    c. Remains the same

    d. None of these

27. There are _____ grains of water vapor per pound of water.

    a. .075

    b. 13.33

    c. 3.42

    d. 7000

28. There are _____ pounds per cubic foot of air.

    a. .075

    b. 13.33

    c. 3.42

    d. 7000

Airflow Tech Tip

The three common methods of duct system sizing are the equal friction method, the extended plenum method and the static regain method.

29. The cfm per ton rate for normal comfort cooling is approximately:

    a. 250 cfm/ton
    b. 300 cfm/ton
    c. 400 cfm/ton
    d. 500 cfm/ton

30. Cooling air raises its:

    a. Specific humidity
    b. Relative humidity
    c. Dew point
    d. None of these

31. Air with a dry bulb temperature of 60 degrees and a wet bulb temperature of 60 degrees has a relative humidity of:

    a. 0%
    b. 50%
    c. 100%
    d. The answer cannot be determined without a psychrometric chart

32. Air with a dry bulb temperature of 60 degrees and a wet bulb temperature of 60 degrees has a dew point temperature of:

    a. 50 degrees
    b. 60 degrees
    c. 70 degrees
    d. The answer cannot be determined without a psychrometric chart

33. A sensible heat factor of 1.0 means that _____ are (is) being removed.

    a. Grains

    b. Water vapor

    c. Moisture

    d. None of these

34. Decreasing the CFM over an evaporator will _____ the moisture removal from the air.

    a. Increase

    b. Decrease

    c. Not change

    d. None of these

35. A rotating Velometer measures:

    a. CFM

    b. FPM

    c. RPM

    d. Feet

36. Return air grilles are not to be installed in:

    a. Bedrooms

    b. Bathrooms or kitchens

    c. Hallways

    d. Living rooms

Motor Tech Tip

Oil provides a cushion between a motor's shaft and bearing. Theoretically, the shaft and bearing do not come in contact because of the layer of oil between them. When the lubrication fails, metal-to-metal contact occurs, the bearing wears and eventually the motor fails.

Airflow Tech Tip

A well-constructed and properly installed duct system may leak as much as 30% of its design airflow through unsealed sheet metal joints. Sealing duct seams can make a substantial difference in system operation, comfort and economy.

37. The proper amount of air in CFM for each room is determined by:

   a. Rules of thumb

   b. A detailed room-by-room heat load calculation

   c. The size of the ducts

   d. Local building code

38. On residential duct installations the accepted method of attaching duct sections to each other is:

   a. Welding them

   b. Using the slip and drive method

   c. The bolted flange method

   d. Slip fit with a suitable sealant

39. Which of the following is not an acceptable location for a thermostat?

   a. Five feet off the floor on an inside wall

   b. Five feet off the floor on an outside wall

   c. Near a return air register

   d. In the most central location of the home

40. What is the difference between "accuracy" and "precision" when applied to instruments?

   a. Accuracy means the instrument gets the same reading each time it is used to measure the same value. Precision means the instrument is correctly calibrated.

   b. Accuracy is a measure of how close the measurement is to being correct. Precision is the measure of how small a value the instrument can detect.

   c. Accuracy is a measure of how small a value the instrument can detect. Precision is a measure of the ability of the instrument to obtain the same reading when taking the reading a second or third time.

   d. Accuracy and precision mean exactly the same thing.

41. The range of an instrument is:

    a. The difference between the lowest and highest readings the instrument can measure

    b. The smallest increment of measurement the instrument is capable of measuring

    c. The difference between the cut-in and cut-out

    d. The highest value the instrument can measure without damage to the instrument

42. A duct with an inside area of 4 square feet has a velocity of 1000 feet per minute. How many CFM of air are moving through that section of the duct?

    a. 1000 CFM

    b. 2000 CFM

    c. 4000 CFM

    d. 6000 CFM

43. A 32-inch by 32-inch square duct is to be changed to a height of 28 inches so it will fit around an obstruction in a boiler room. How wide must the duct be in order to maintain the same area? Round the number off to the nearest whole number of inches.

    a. 36 inches

    b. 37 inches

    c. 38 inches

    d. 39 inches

44. Which of the following is not a sheet metal duct fitting?

    a. Register boot

    b. End boot

    c. 90 degree half union

    d. Starting collar

Power Transmission Tech Tip

A "keyway" is a groove in the shaft and hub of a sheave or pulley. A "key" is a piece of rectangular steel that fits into the keyway thus locking the shaft and hub together to prevent slippage.

Notes

45. Which of the following is a true statement?

    a. The greater the velocity, the more resistance there is to a change in direction

    b. If the duct area decreases, the velocity increases

    c. When a fitting creates turbulence, the turbulence continues for several feet down the duct

    d. All of these are correct statements

46. Which of the following is not a duct system type?

    a. Radial duct system

    b. Parameter duct system

    c. Extended plenum system

    d. Equal friction system

47. A 24-inch by 8-inch rectangular duct has an aspect ratio of:

    a. 1:1

    b. 2:1

    c. 3:1

    d. 4:1

48. The most efficient method of air balancing a duct system is:

    a. The equal temperature method

    b. The sequential method

    c. The proportional method

    d. The flow hood method

49. Which of the following types of turning vanes produce the least dynamic system loss?

    a. Single wall vanes

    b. Double wall vanes

    c. Heel and cheek vanes

    d. Not using turning vanes at all

50. Which of the following equations is used to determine the actual operating capacity of a system?

    a. BTUH = 1.08 × CFM × TD between the return and supply air

    b. BTUH = 4.5 × CFM × The difference in enthalpy between the return and supply air

    c. BTU = specific heat of air × pounds of air × TD

    d. None of these

Notes

Notes

PART TEN
Controls

Controls Tech Tip

Discharge any static electrical charge that may have built up on your body before handling electronic circuit boards or components. Before handling electronic components touch a large metal object to discharge any stray charges from your body.

The controls industry is a specialty within the HVAC industry. Many HVAC technicians and mechanics are specialists in controls and the application of controls to heating, ventilating, air conditioning and refrigeration. Considering the history of the heating and cooling business, controls as a specialty career choice is a relatively new entrant to the HVAC business. However, the application of controls to heating and cooling equipment has grown in scope, application and sophistication to the extent that it has become an important specialty within the HVAC industry. It is often stated by experts that about 60% of commercial and business buildings use 40% or more energy than necessary. Heating, cooling and ventilation equipment consume the vast majority of the energy required to operate a typical building.

Although controls is a specialty with its own industry certification, everyone working in the HVAC industry deals with controls to one degree or another and therefore must have some knowledge and understanding of controls. It is simply not possible to work as a mechanic, technician or building operator without some knowledge of controls. Therefore, controls questions appear to some degree on nearly every heating or cooling exam to one degree or another. For example, heat pumps rely on extensively on controls as do economizers and energy management systems.

The topic of controls can be subdivided into three major categories; 1) Pneumatic 2) Electric and 3) Electronic. With the general growth and acceptance of personal computers electronic controls have expanded exponentially over the past 20-years or so.

The following 50-question multiple choice practice exam contains a good sampling of the more common control questions likely to appear on many HVAC exams including general air conditioning exams, heat pump exams, gas heat and oil heat tests as well as control specialty exams. Those preparing to sit for an exam that specifically concentrates on controls or a control certification are encouraged to augment this study section with additional preparation and study prior to taking a controls specialty exam.

CONTROLS PRACTICE EXAM

1. Pneumatic controls utilize _____ power for their operation.

   a. Air

   b. Water

   c. Electrical

   d. Solar

2. Setpoint is:

   a. The desired condition

   b. The current condition

   c. The difference between cut-in and cut-out

   d. The very center of the range

3. Given the following which of the following is the most utilized control transformer size?

   a. 10VA

   b. 40VA

   c. 100VA

   d. 120VA

4. The difference between cut-in and cut-out is the:

   a. Differential

   b. Range

   c. Throttling range

   d. Setpoint

5. DDC stands for:

   a. Direct Dial Connection

   b. Direct Digital Control

   c. Dual Digital Controls

   d. Digital Dial Connection

Controls Tech Tip

Passive sensors do not require a power supply to operate while active sensors require a source of power such as an air or electrical supply. An example of a passive sensor is a bimetal strip and an example of an active sensor is a powered smoke detector.

Controls Tech Tip

The function of a sensor is to detect a physical change and create a corresponding signal representative of the change.

6. A 20VA control transformer rated for 24-volts on the secondary can safely draw a maximum secondary amperage of:

    a. .50

    b. .83

    c. 1.25

    d. 8.0

7. A control that opens its contacts upon a rise in temperature or pressure is a:

    a. Direct acting control

    b. Reverse acting control

    c. Dual acting control

    d. Thermistor control

8. On electro-mechanical controls which of the following control settings is typically not adjustable?

    a. Range

    b. Setpoint

    c. Differential

    d. Cut-out

9. Control "status" is an:

    a. Analog input

    b. Analog output

    c. Digital input

    d. Digital output

10. Which of the following is an example of a digital output?

    a. Fan motor status

    b. Pump command

    c. Smoke alarm

    d. VFD speed control

11. Which of the following is an example of an analog input?

    a. Temperature sensor

    b. Fan motor status

    c. Smoke alarm

    d. Pump command

12. What type of heat does a thermistor measure?

    a. Sensible heat

    b. Latent heat

    c. Specific heat

    d. Non-linear heat

13. Using a computer to track and graph temperatures over a period of time is called:

    a. Totalization

    b. Trending

    c. Graphics

    d. Auditing

14. Using a computer to track and record the consumption of electrical power is called:

    a. Totalization

    b. Trending

    c. Graphics

    d. Auditing

Controls Tech Tip

Binary relates to a choice between two states, on or off, 1 or 0, open or closed. Analog relates to a variable anywhere between two extremes such as a changing temperature or the position of a damper.

Controls Tech Tip

Status is an input from a device which detects if a motor or other similar device is operating. A typical status switch for a motor is a current sensing relay. If the motor is operating the current to the motor is detected and the relay operates a switch contact which becomes an input to a controller informing the controller that the motor is operating.

15. Which of the following discrete controls is most likely to be utilized on an airside economizer?

    a. Pneumatic relay

    b. Enthalpy control

    c. High pressure control

    d. Low pressure control

16. BACnet is:

    a. A protocol established by ASHRAE

    b. A protocol established by ARI

    c. A controls protocol owned by RSES

    d. A controls protocol owned by Johnson Controls Inc.

17. An "ASC" is:

    a. An Application Specific Control

    b. An Anti Shortcycling Controller

    c. An ASHRAE Specified Controller

    d. An Automated Specialty Controller

18. A "global point" is:

    a. A point shared between controllers

    b. A series of points connected so as to communicate to each other

    c. Always used to average a series of temperature readings

    d. Not a valid control term

19. An incremental control actuator motor typically uses _____ wires.

    a. two

    b. three

    c. four

    d. six

20. Which of the following would best be described as an "actuator"?

    a. A damper motor

    b. A damper

    c. A steam valve

    d. A temperature sensor

21. "Demand Ventilation" is a control strategy utilizing:

    a. An oxygen sensor

    b. A carbon dioxide sensor

    c. A carbon monoxide sensor

    d. An enthalpy control

22. A "duct static pressure sensor" is typically used in a:

    a. Constant volume duct system

    b. Variable air volume duct system

    c. Single pass duct system

    d. Demand ventilation system

23. An "EP" switch is:

    a. An electrically operated pneumatic switch

    b. A pneumatically operated electric switch

    c. An electronic-pneumatic receiver controller

    d. A pneumatic-electronic receiver controller

24. A "PE" switch is:

    a. An electrically operated pneumatic switch

    b. A pneumatically operated electric switch

    c. An electronic-pneumatic receiver controller

    d. A pneumatic-electronic receiver controller

Controls Tech Tip

A digital output command to a motor and the return status input to the controller must be synchronized so that the controller can expect a status input that the motor is operating only if the motor is commanded to operate. A status that the motor is operating when the motor is off indicates a problem just as no status indication of a running motor when commanded on indicates a problem.

Controls Tech Tip

Most DDC control systems have an alarming indication function to inform the operator of the existence of an abnormal operating condition.

25. A "supervisory" control is:

    a. a control which can override other controls under its authority

    b. a control which resides on a distributed network or system

    c a control which often contains algorithms too complex for local control

    d. all of the above are correct

26. The three parts of a control system are:

    a. Sensor, Controller & Actuator

    b. Sensor, Controller & Power assembly

    c. Sensor, Controller & Damper

    d. Sensor, Data Link & Computer

27. An "open loop" control system:

    a. Has feedback

    b. Has no feedback

    c. Is powered pneumatically

    d. Is powered electrically

28. An example of a "controlled variable" is:

    a. Airflow

    b. A damper

    c. 0 to 10 VDC

    d. A setpoint

29. An example of a "controlled device" is:

    a. A damper

    b. Water flow

    c. A humidity sensor

    d. An enthalpy control

30. A 'transducer':

   a. Converts one form of energy to another

   b. Serves as an AC to DC function generator

   c. Changes pneumatic pressures to graphical language

   d. Is the final element in the control process

31. A "varying signal" can best be described as a:

   a. Modulating signal

   b. Discrete signal

   c. Digital signal

   d. Feedback signal

32. A modulating control output can also be described as a:

   a. Proportional signal

   b. Filtered signal

   c. Branch input

   d. Reverse acting output

33. The _____ is the range at which the temperature has been satisfied and there is no demand or call for heating or cooling.

   a. deadband

   b. setpoint

   c. differential

   d. aspect ratio

34. The "normal" position of a set of contacts in a control relay are in their normal position when:

   a. The relay is de-energized

   b. The relay is energized

   c. The relay is in its normal condition

   d. None of the above are correct

Controls Tech Tip

Networked DDC control systems may also connect to the Internet allowing access to the HVAC equipment control system from anywhere where the Internet is available.

Controls Tech Tip

Networked DDC control systems usually have access control security thru the use of assigned usernames and passwords and most also use firewalls to prevent unauthorized access.

35. RAM memory:

    a. Maintains its contents when the power is shut down

    b. Loses its contents when the power is shut down

    c. Can only be written to once

    d. Is a sector of memory located on the hard-drive

36. Which of the following applications of an actuator is most apt to require a long operating stem?

    a. A damper application

    b. A steam valve application

    c. A hot water valve application

    d. A safety application

37. In the "unoccupied mode" the outside air intake damper is typically:

    a. Wide open

    b. Closed

    c. Allowed to modulate

    d. Center positioned

38. The core of a building typically:

    a. Requires heating year-round

    b. Requires cooling year-round

    c. Requires an equal amount of heating & cooling

    d. Requires a higher degree of air filtration

39. The perimeter of a building generally:

    a. Has greater heat losses than the core

    b. Has greater heat gains than the core

    c. Has greater heat gains and losses than the core

    d. Has the least need for heating or cooling

40. The rate at which control system data is transferred is measured in:

    a. Baud

    b. RAM

    c. EEPROM

    d. Hexadecimal Rate

Controls Tech Tip

Some knowledge of general electronics and electronic concepts can be very useful when working with DDC controls applied to HVAC systems.

41. A common connection point for devices on a network is called:

    a. A hub

    b. A gateway

    c. A controller

    d. A repeater

42. A common set of rules or standards that governs the exchange of information over a digital communications system is called:

    a. A protocol

    b. A presentation layer

    c. An application entity

    d. Enumeration

43. Which of the following devices performs the function of changing or translating data from one form to another and back again?

    a. A router

    b. A hub

    c. A repeater

    d. A bridge

44. A group of two or more computers linked together is called:

    a. A network

    b. A bridge

    c. A hub connection

    d. A bus

Controls Tech Tip

Indoor and outdoor lighting loads may use 20% or more of the power consumed in commercial buildings and are therefore good candidates for energy conservation. Light sensor controls and scheduling are two effective strategies for conserving energy.

45. A networked controller communicates over a set of wires known as a:

    a. Communications bus

    b. Communications hub

    c. Gateway

    d. Tunnel

46. The difference between the setpoint and the actual value of the controlled variable is known as:

    a. Differential

    b. Deviation

    c. Compensation control

    d. Integral action

47. A device that converts a sensor signal to an input signal that is usable by a controller is called a:

    a. Transmitter

    b. Router

    c. Bridge

    d. Sensor

48. A sensor with a positive temperature coefficient:

    a. Increases in resistance with an increase in temperature

    b. Decreases in resistance with an increase in temperature

    c. Opens its contacts on a temperature rise

    d. Closes its contacts on a temperature rise

49. One method of controlling overshoot is the use of:

    a. A cooling compensator

    b. A heating anticipator

    c. A ratio resistor

    d. A silicon controlled rectifier

50. A remote or local controller located near the HVAC equipment usually allows for _____ operation should a communications bus failure occur.

    a. standalone

    b. alarm

    c. differential

    d. integral

Controls Tech Tip

A "footcandle" is the amount of light that is produced by a lamp in lumens divided by the area that is illuminated.

Notes

Proportional gain is the amount of output change per unit of input change sometimes called sensitivity.

PART ELEVEN
Indoor Air Quality

Indoor Air Quality Tip

PPM stands for parts per million. Given a count of one million particles one ppm is one particle out of the total of one million. Many substances contained in air are measured in parts per million.

Several testing organizations have established technician certifications in the area of indoor air quality. Additionally, many of the other heating, ventilating, air conditioning and comfort control certifications and contractor licensing exams contain questions on the exams regarding indoor air quality. In particular, an important aspect of indoor air quality is the safety issue related to carbon monoxide.

Regardless of which HVAC exam you may be sitting for you would do well to be prepared for at least several questions regarding indoor air quality and/or carbon monoxide. In this test prep section are 50 of the most commonly asked questions about indoor air quality and carbon monoxide. Mastering these questions will provide a strong foundation for answering most of what appears on indoor air quality sections of a high percentage of technician certification exams.

INDOOR AIR QUALITY EXAM

1. Which of the following is known as the "silent killer"?

    a. Carbon dioxide

    b. Carbon tetrachloride

    c. Carbon monoxide

    d. Monodioxide

2. _____ ventilation is a control strategy used to maintain a high oxygen content in buildings where a healthy environment is of particular concern.

    a. Demand

    b. Outside

    c. Purge

    d. Smoke

3. The most economical means of measuring the amount of oxygen deprivation in an occupied space is thru the use of:

    a. A carbon dioxide sensor

    b. A oxygen sensor

    c. A nitrogen sensor

    d. A smoke detector

4. Which of the following is the natural result of combustion?

    a. Carbon dioxide

    b. Water vapor

    c. Carbon monoxide

    d. All of the above

Indoor Air Quality Tip

Indoor air quality is an important and growing aspect of the heating, ventilating, air conditioning and environmental control industry. Anyone working in the HVAC industry needs to be aware of the issues and responsibilities surrounding indoor air quality.

Indoor Air Quality Tip

Indoor air quality issues have resulted in costly legal actions against building owners, those who lease buildings, the building maintenance organization and outside contractors. Anyone involved with the operation of a suspect building may be legally bound and liable for indoor air quality problems.

5. The most serious result of incomplete combustion in any fuel burning heating device is:

   a. Carbon dioxide

   b. Water vapor

   c. Carbon monoxide

   d. Nitrogen

6. Carbon monoxide poisoning is often mistaken for:

   a. The flu with a headache

   b. Cancer

   c. A mild cold

   d. None of the above

7. Carbon monoxide is measured in:

   a. Microns

   b. PPM

   c. Inches of water gage

   d. Percent

8. Outdoor air intake dampers should be located away from:

   a. Busy ground level intersections

   b. Loading docks

   c. Sewer vents

   d. All of the above

9. "Infiltration" is:

   a. Air that enters a space without filtration or control

   b. Air that enters a space without filtration

   c. Air that leaves a building thru cracks and unchecked door seals

   d. None of the above

10. "Offgasing" is:

    a. The release of volatile compounds from products and construction materials

    b. Usually harmful compounds and chemicals which slowly enter living spaces over time

    c. Chemicals released into the air especially as the host material is heated

    d. All of the above

11. In the science of indoor air quality VOC stands for:

    a. Vocational-Occupational Contaminants

    b. Volatile Occupational Contaminants

    c. Volatile Offgasing Contaminants

    d. Volatile Organic Compounds

12. Which of the following are common symptoms of carbon monoxide poisoning?

    a. Headache & nausea

    b. Fatigue & dizziness

    c. Shortness of breath

    d. All of the above

13. Nitrogen dioxide is a toxic resulting from:

    a. The process of combustion

    b. Rapid oxidation

    c. Both a and b are correct

    d. Neither a nor b

Indoor Air Quality Tip

State and local codes often adopt ASHRAE indoor air quality standards into the regulations at which time the standards become enforceable as law.

Indoor Air Quality Tip

Legal action regarding indoor air quality issues can be brought to a court of law even if the state or local regulatory code agencies do not have a specific code addressing air quality.

14. The quality of air indoors can often be worse than that of the outdoors.

    a. T

    b. F

15. "Mitigation" means:

    a. To alleviate

    b. To increase

    c. To take legal action

    d. To document

16. Solid particles suspended in air are measured in:

    a. Microns

    b. Parts per million

    c. Milligrams

    d. Hundredths of an inch

17. Which of the following organizations has legal oversight over air quality in the United States?

    a. EPA

    b. ARI

    c. RSES

    d. ASHRAE

18. The higher the velocity of the air thru a fiber filter the:

    a. More effective the filtration

    b. The less effective the filtration

    c. Air velocity had no affect on filtration

    d. None of the above is correct

19. At what level of indoor air carbon monoxide do most fire departments require emergency personnel to wear self contained breathing gear (SCBA)?

    a. 8 ppm

    b. 35 ppm

    c. 50 ppm

    d. 100 ppm

20. The function of an ERV is to:

    a. Exchange heat between exhaust and intake air

    b. Control the static pressure in the main trunk duct

    c. Control the supply water pressure in high-rise buildings

    d. Prevent the accumulation of VOCs in the fresh air intake

21. ASHRAE standard 55-1992 specifies the conditions likely to be thermally acceptable to _____ percent of adult occupants of a conditioned space.

    a. 50%

    b. 60%

    c. 75%

    d. 80%

22. Indoor air quality is legally the responsibility of:

    a. The building owner

    b. The business leasing the building

    c. Those operating and maintaining the building

    d. All of the above

Indoor Air Quality Tip

Measuring the amount of carbon dioxide in indoor air as an indicator of the amount of oxygen present is often more economical than measuring the oxygen content directly.

Indoor Air Quality Tip

Do not confuse carbon dioxide with carbon monoxide. Carbon monoxide is a deadly poison while carbon dioxide is not.

23. Clean rooms are classified according to:

    a. The particle count per cubic foot of air

    b. Microns allowed per square foot of surface area

    c. Parts per million allowed per cubic centimeter

    d. Percent of filtration at 700 fpm of linear airflow

24. HEPA stands for:

    a. High Efficiency Particulate Air filter

    b. High Efficiency & Performance Air filter

    c. Highly Effective Performance Air filtration

    d. Heating Efficiency Performance Association

25. The internal static pressure in a building should be approximately:

    a. .5 inches higher than the outside air pressure

    b. .05 inches higher than the outside air pressure

    c. .5 inches lower than the outside air pressure

    d. .05 inches lower than the outside air pressure

26. Activated charcoal is often satisfactory for the removal of:

    a. Gas-phase toxic or odorous pollutants

    b. Particulates of .5 microns or larger

    c. Particulates of 12 microns or larger

    d. Fibrous insulation suspended in air

27. A _____ biological safety cabinet is a gastight negative pressure containment system that provides a physical barrier between the agent and the worker.

    a. Class 1

    b. Class 2a

    c. Class 2b

    d. Class 3

28. Is refrigerant oil that has gone thru a compressor burn-out classified as a toxic waste?

    a. Yes, if the oil contains a substance classified by the EPA as a toxic waste

    b. Yes, because it contains acids

    c. Yes, the EPA classifies all substances that have burned as toxic waste

    d. No, none of the above is correct

29. Three common methods used to mitigate poor indoor air quality are:

    a. Filtration, ventilation and exhaust

    b. Filtration, capture and exhaust

    c. Capture, removal and reduction

    d. Removal, reduction and retention

30. A micron is:

    a. One 1,000th of an inch

    b. One 10,000th of an inch

    c. One 25,400th of an inch

    d. One millionth of an inch

31. Bacteria, molds, pollen, and viruses are types of:

    a. Chemical contaminants

    b. Biological contaminants

    c. Combustion contaminants

    d. Physical contaminants

Indoor Air Quality Tip

Often there is a fine balance between maintaining good indoor air quality and conserving energy. Indoor air quality must be carefully considered when energy conservation strategies are considered.

Indoor Air Quality Tip

Carbon monoxide is present in the exhaust of all combustion systems including but not limited to BBQ grills, gas and oil furnaces and boilers as well as truck and auto exhaust. These and other sources of combustion gases must not be allowed to enter living spaces.

32. According to the World Health Organization approximately what percentage of buildings suffers from SBS (Sick Building Syndrome)?

    a. 10%

    b. 15%

    c. 25%

    d. 30%

33. A key factor in solving indoor air quality problems is to:

    a. Carefully measure the types and quantity of contaminants

    b. Locate and isolate the source of the pollution

    c. Ensure all occupants are aware of the problem

    d. Keep exact records of the actions taken

34. Which of the following steps would be the first step in the process of investigating an indoor air quality problem at a building?

    a. Conducting an IAQ survey

    b. Reviewing the HVAC system type and condition

    c. Performing a full-scale IAQ assessment

    d. Defining the perceived problem

35. Physical stressors are those factors which:

    a. Can be sensed, felt, seen or heard

    b. Make the occupant uncomfortable

    c. May include factors the occupant is not aware of

    d. All of the above

36. Sick Building Syndrome is highly suspected when:

    a. 20% or more of the building occupants complain with physical symptoms within a two-week period

    b. A high degree of occupants call in sick with symptoms of headaches, fatigue, dizziness, coughing, nausea and eye or throat irritations

    c. A high degree of occupants find that their symptoms are alleviated by leaving the building

    d. All of the above

37. During an initial pre-investigatory building walk-thru which of the following measurements may be taken?

    a. Nitrogen dioxide, carbon monoxide and tobacco smoke readings

    b. Volatile compounds, oxides and radon measurements

    c. Readings for particulate count and biological sampling

    d. Carbon dioxide, temperature and relative humidity readings

38. Which of the following gases lead the list as originating from man-made materials within a building?

    a. Nitrous oxide

    b. Carbon monoxide

    c. Nitrogen dioxide

    d. Formaldehyde

39. Which of the following are psychosocial factors affecting personal comfort:

    a. The presence of residual tobacco smoke in the building

    b. The presence of carbon dioxide in the air

    c. The amount of personal work space and quality of furnishings

    d. None of the above

Indoor Air Quality Tip

Tighter highly insulated energy efficient modern buildings conserve a great deal of energy but then create the potential for indoor air quality problems. Pollutants that enter the indoor air are more likely to be trapped in the building unless collected or exhausted to the outdoors.

Indoor Air Quality Tip

Just as traps under sinks and sewers prevent sewer gasses from entering living spaces, condensate traps on packaged rooftop systems need traps to prevent untreated air as well as insects and rodents from entering the air delivery system. Dry condensate traps also allow air, insects and rodents to enter the equipment.

40. Which of the following can affect the intensity of a person's reaction to an irritant in the air?

    a. Temperature

    b. Humidity

    c. Temperature and humidity

    d. Presence of others in the area

41. A fairly common but not regularly recognized source of poor indoor air quality and increased dust and residue in homes is:

    a. The use of candles

    b. Synthetic carpets

    c. Down filled comforters

    d. Computer monitors

42. Which of the following substances has been directly attributed to as many as 20,000 cases of lung cancer in the United States?

    a. Radon gas

    b. The use of candles

    c. High levels of carbon dioxide

    d. High levels of nitrogen

43. Concentrations of radon gas is measured in:

    a. Picocuries

    b. Microns

    c. Ppm

    d. Micrograms

44. Particulates are categorized in terms of:

    a. Size and source

    b. Size and density

    c. Density and diameter

    d. Size and color

45. Excessively high concentrations of carbon dioxide in indoor air is usually an indication of:

    a. A lack of ventilation air

    b. An increased rate of infiltration air

    c. SBS

    d. Tobacco smoke in the air

46. Which of the following IAQ problems was first discovered to have it source in HVAC cooling towers?

    a. Legionnella

    b. Poor air distribution

    c. Inadequate makeup air

    d. Radon

47. Organic compounds are:

    a. Compounds which contain carbon

    b. Compounds which contain synthetics

    c. Compounds consisting of radioactive elements

    d. Those which can only be created in a laboratory

Indoor Air Quality Tip

The use of less expensive prefilters in front of expensive bag filters improves indoor air quality and extends the longevity of the expensive filters.

Indoor Air Quality Tip

Some knowledge of the fundamentals of chemistry and microbiology is useful in understanding and dealing with indoor air quality issues.

48. A compound is:

    a. A material made up of two or more elements electrically bound to one another.

    b. A material made up of three or more elements chemically bound to one another.

    c. A material made up of excess electrons

    d. A naturally occurring element

49. TLV stands for:

    a. Threshold Limit Value

    b. Threshold Limit Vortex

    c. Theoretical Limit Value

    d. Theoretical Lowest Value

50. The average urban outdoor air concentration of carbon dioxide has been determined to be:

    a. 40 ppm

    b. 400 ppm

    c. 4000 ppm

    d. 40,000 ppm

PART TWELVE
Final Practice Exams

- A Short 50 Question Final Practice Exam

- A Full-Length 100 Question Final Practice Exam

Notes

If you have been carefully using this study guide and taking the practice exams by chapter, you should be well prepared for these two final practice exams. If you are comfortable with the material in the previous chapters, then you are ready to take up the challenge of these two exams.

Anyone able to achieve 70% or higher on these two exams is well prepared to pass either the NATE or the HVAC Excellence exams. Much of the material and practice exams in this book will also serve as a good foundation for taking the more difficult RSES CM and CMS exams. Mastering the NATE and/or HVAC Excellence exams proves you are ready for the RSES CM and CMS challenges.

One authority has made the claim that NATE certification is like having a bachelor's degree in HVAC, RSES CM certification is like having a master's degree and RSES CMS certification is like having a doctorate in HVAC.

Accordingly, the industry is quickly recognizing technician certifications and in some cases companies are requiring employees to become certified. With the aid of this book you will be in a far better position to successfully pass whichever exam you wish to take.

SHORT FINAL PRACTICE EXAM

1. Generally, vibration eliminators in piping should be installed:

   a. At a 45 degree angle

   b. At a 90 degree angle

   c. Perpendicular to the compressor crankshaft

   d. Parallel to the compressor crankshaft

2. If you purchased ten filter-driers for $100.00, what was the price of each filter-drier?

   a. $100.00

   b. $ 10.00

   c. $ 1.00

   d. $ 1.10

3. Which of the following jobs requires safety glasses?

   a. Driving a vehicle

   b. Extending a ladder

   c. Brazing

   d. Pulling thermostat wire

4. Which of the following instruments would typically be used to measure a vacuum?

   a. Micron gauge

   b. Thermostat

   c. Anemometer

   d. Differential pressure gauge

5. ASHRAE Standard 34 deals with:

   a. Carbon monoxide safety levels

   b. Refrigerant quality

   c. Equipment room ventilation and alarms

   d. Refrigerant toxicity and flammability

6. What control would you use on a capillary tube or automatic expansion valve system to maintain the desired temperature?

   a. Dual pressure control

   b. High pressure control

   c. Thermostatic temperature control

   d. Low pressure control

7. The total heat contained in any substance is obtained by adding together its:

   a. Radiant and specific heat

   b. Latent and sensible heat

   c. Radiant and sensible heat

   d. Specific and latent heat

8. An outdoor air conditioning condensing unit should not be located:

   a. In the sun

   b. On the side of the house

   c. Within 5 feet of a hose bib connection

   d. Where any objects or obstructions are within 5 feet of the condenser fan

9. An outdoor air conditioning condensing unit should not be located:

    a. Under an overhang where water may drain onto it

    b. Within 4 feet of a hose bib connection

    c. Within 3 feet of an electrical disconnect

    d. All of the above

10. A thermostat should be mounted:

    a. Approximately 5 feet above the floor

    b. Level

    c. On an inside wall

    d. All of the above

11. As the pressure on a liquid rises, the temperature at which the liquid will boil:

    a. Varies

    b. Remains the same

    c. Increases

    d. Decreases

12. A voltmeter is connected to a circuit:

    a. In series

    b. In parallel

    c. Crossover

    d. Delta

13. Pressure relief valves are not to be piped:

    a. In parallel

    b. In series

    c. Facing up

    d. Within 12" of a wall

Notes

14. Which of the following is typically covered by local codes?

   a. Refrigerant type

   b. Maximum length of refrigerant lines

   c. Location of condensing unit disconnect

   d. Location of liquid and suction line service valves

15. Hard drawn ACR tubing may not be used:

   a. In combination with soft drawn ACR tubing

   b. Without first annealing the tubing

   c. With a swedge joint

   d. On residential applications

16. Which of the following oils is the most hygroscopic?

   a. POE oil

   b. Mineral oil

   c. Vegetable oil

   d. Vacuum pump oil

17. Highly hygroscopic oils must be kept in:

   a. Metal or glass containers

   b. Tightly capped plastic containers

   c. An area out of direct sunlight

   d. A refrigerated space 50 degrees or cooler

18. The purpose of technician certification is to:

   a. Raise the competency level of technicians

   b. Decrease the number of poorly installed and improperly serviced systems

   c. Help consumers choose well-trained technicians

   d. All of the above

19. The purpose of converting psig to psia prior to calculating the compression ratio of a system is to:

    a. Make the answer come out as a positive number if the low side pressure is in a vacuum

    b. Make sure the answer is never negative

    c. Make the answer mathematically correct

    d. Allow for easy conversion to the metric system

20. As a warm air furnace raises the temperature of the air in a structure, the relative humidity of the space will:

    a. Increase

    b. Decrease

    c. Increase by the square of the change in temperature

    d. Remain the same

21. ACR soft copper tubing comes in:

    a. 20 foot lengths

    b. 30 foot sealed rolls

    c. 50 foot rolls

    d. 50 foot sealed rolls

22. Typical ACR copper tubing flares are:

    a. Double flares

    b. 33 degree flares

    c. 45 degree flares

    d. 60 degree flares

Notes

23. Customers generally trust technicians who:

    a. Look them straight in the eye during conversations

    b. Genuinely listen to the customer's concerns

    c. Provide them with accurate information and options

    d. All of the above are correct

24. What amount of oxygen is normally contained in air?

    a. 10%

    b. 15%

    c. 21%

    d. 25%

25. The "critical" temperature and pressure of a refrigerant is:

    a. The conditions above which a refrigerant gas cannot be condensed into a liquid

    b. The conditions below which a refrigerant cannot be vaporized to a gas

    c. The conditions where even if a change of state were to occur, no latent heat exists

    d. The pressure at which no heat exists in the refrigerant therefore the temperature is absolute zero

26. Refrigerant cylinders should never be filled to over:

    a. 50%

    b. 65%

    c. 80%

    d. 90%

27. Which of the following gases should never be used to pressure test for leaks?

    a. Nitrogen

    b. Oxygen

    c. Carbon dioxide

    d. None of these should be used for pressure testing

28. What is the nominal operating wattage of a one-quarter horsepower electric motor?

    a. 186 watts

    b. 100 watts

    c. 46 watts

    d. 373 watts

29. The electrical resistance of a thermistor _____ with a temperature increase.

    a. increases

    b. decreases

    c. remains unchanged

    d. varies by the 10th power

30. Removing refrigerant from a system and storing it in a cylinder without performing any testing or processing is called:

    a. Recycling

    b. Reclaiming

    c. Recovering

    d. Restoring

Notes

31. Processing used refrigerant to bring it up to the standards of new virgin refrigerant is called:

    a. Recycling

    b. Reclaiming

    c. Recovering

    d. Restoring

32. The quality standard for new virgin refrigerant is:

    a. ARI 700

    b. ARI 740

    c. ARI 2030

    d. UL 15

33. Which of the following is not a suitable control for an air side economizer as used on a packaged unit air conditioning system?

    a. Enthalpy control

    b. Outside air thermostat

    c. Neither a nor b

    d. Low pressure control

34. What two pressures can a pitot tube measure?

    a. Static and total

    b. Velocity and total

    c. Velocity and static

    d. Total and external static pressure

35. An oil separator has which of the following lines or connections attached to it?

    a. Inlet, outlet, oil line

    b. Inlet and outlet

    c. Inlet and oil line

    d. None of these is correct

36. A three-position service valve located on the outlet of a receiver should be left in what position for normal operation with no gauges attached?

    a. Front-seated

    b. Mid-seated

    c. Back-seated

    d. Cracked

37. A cooling tower is treated for what two conditions?

    a. Algae and scale

    b. Algae and lime

    c. Scale and disease

    d. Algae and disease

38. Which of the following documents are required by law?

    a. Air balance report

    b. Charging chart

    c. Equipment list

    d. MSDS sheets

Notes

39. A heating and/or cooling load calculation is used to:

    a. Choose the correct equipment size

    b. Determine the correct duct sizes

    c. Determine the correct cfm to each room

    d. All of the above

40. What component of a dehumidifier can be used to reheat the air after dehumidification has taken place?

    a. Evaporator

    b. Condenser

    c. Compressor

    d. None of the above

41. Type 1 EPA refrigerant certification is for systems:

    a. Containing 5 pounds or less of refrigerant

    b. Containing 8 pounds or less of refrigerant

    c. Containing 5 pounds or less of refrigerant and having fixed service valves

    d. Containing 5 pounds or less of refrigerant and having no service valves

42. The operating pressures in an R-410A system are:

    a. 40% to 70% higher than in an R-22 system

    b. 10% to 15% higher than in an R-22 system

    c. Lower than in an R-22 system

    d. Approximately the same as in an R-22 system

43. R-410A systems use which of the following refrigerant oils?

    a. Mineral

    b. Alkylbenzene

    c. Polyol ester

    d. Glycols

44. Carbon monoxide poisoning is often mistaken for:

    a. The flu

    b. Diabetes

    c. Blood disease

    d. Chronic low blood pressure

45. Carbon monoxide is the result of:

    a. Incomplete combustion

    b. Lowering the temperature of the flame

    c. Excessive fuel or a lack of oxygen

    d. All of the above

46. The complete combustion of natural gas requires:

    a. 5 cubic feet of air per cubic foot of gas

    b. 10 cubic feet of air per cubic foot of gas

    c. 15 cubic feet of air per cubic foot of gas

    d. 20 cubic feet of air per pound of natural gas

Notes

47. A tightly constructed building generally causes the pressure inside the building to be _____ than the pressure outside.

    a. Higher

    b. Lower

    c. The same

    d. None of these

48. R-410A is a _____ refrigerant.

    a. CFC

    b. HCFC

    c. HFC

    d. POE

49. A refrigerant classified as an "Azeotropic blend":

    a. Will not fractionate

    b. Behaves much like a single-component pure refrigerant

    c. Does not have a temperature glide

    d. All of the above are correct

50. Superheat is:

    a. The difference between the saturation temperature in the evaporator and the temperature of the refrigerant leaving the evaporator

    b. The difference between the saturation temperature in the evaporator and the saturation temperature in the condenser

    c. The difference between the temperature measured at the compressor suction and the compressor discharge lines

    d. A single temperature taken in the center of the evaporator minus the ambient air temperature

NATE stands for North American Technician Excellence, a voluntary set of examinations created to certify that technicians meet an industry standard of technician competency. There is a core exam of 50 questions plus a series of "specialty exams" consisting of 100 questions. For more specific information you may go to the NATE website located at www.natex.org. NATE claims that the exams test for what 80% of technicians should know 80% of the time. NATE also claims that those who took an exam without a review course or seminar were able to pass 45% of the time while 72% passed if they took a review course or seminar prior to taking the exam.

None of the questions in this guide are directly from the actual exams you will take. They are sample questions to give you an idea of what the test is like as well as an aid to review and prepare. Some of the questions are actual NATE questions that have been removed from the exams, retired and placed in the public domain as example questions. The following sample exam contains NATE-like questions that will appear on the core as well as the specialty exams. Like the NATE exams, they are in no particular order. In addition, the NATE exams also contain questions about wiring diagrams. Therefore, it is to your advantage to review and practice reading and understanding HVAC wiring diagrams.

Notes

FULL-LENGTH FINAL PRACTICE EXAM

1. A condensing unit should be mounted on a slab that is level to:

   a. Allow any accumulated water to evenly run out of the cabinet

   b. Increase airflow through the fin area

   c. Prevent the unit from tipping over

   d. Make certain any top discharge air blows straight up

2. A simple time-activated switch is an example of a control that does not use:

   a. Feedback

   b. Electronics

   c. Calibration

   d. Manual adjustment

3. Which of the following systems would typically require a blower with a 2000 cfm output?

   a.   2-ton ac system

   b.   3-ton ac system

   c.   5-ton ac system

   d. 7.5-ton ac system

4. A typical induced-draft furnace uses a blower-off delay in the cooling mode to:

   a. Increase the efficiency of the system

   b. Air wash the coil on shutdown

   c. Minimize dehumidification at the coil

   d. Prevent blowing moisture off the coil on shut off

5. The number 3,412 btu is equal to:

   a. 1 watt

   b. 3.4 watts

   c. 1 kilowatt

   d. 3.4 kilowatts

6. Which of the following would a technician use to check the operating sequence of an air conditioning system?

   a. Schematic

   b. Pictorial diagram

   c. Illustration

   d. Isometric drawing

7. Gas heating systems on roofs or other elevated locations shall be accessible by way of the building if the building is taller than:

   a. 10 feet

   b. 15 feet

   c. 20 feet

   d. 25 feet

8. Wood studs in typical residential construction are how far apart?

   a. 12 inches

   b. 14 inches

   c. 16 inches

   d. 18 inches

9. Which of the following components is energized first after an intermittent ignition gas furnace calls for heat?

   a. Induced draft motor

   b. Pilot valve

   c. Blower motor

   d. Gas valve

10. To have a system automatically switch to heating or cooling as the load changes, the thermostat:

    a. System switch should be in auto

    b. System switch should be in off

    c. System switch should be on

    d. Fan switch should be in auto

11. A typical heat pump uses a high-pressure control to shut off the compressor if the condensing pressure becomes excessive. The control uses a switch that makes the compressor contact on a pressure:

    a. Rise at the compressor discharge

    b. Fall at the compressor discharge

    c. Difference between the high and low side

    d. Combination of the high and low side

12. Which of the following is typically covered by local codes?

    a. Minimum cfm per outlet

    b. Maximum length of a trunk

    c. Termination of condensate line

    d. Refrigerant charge

13. What is the difference between the wiring of an electromechanical thermostat and an electronic thermostat?

    a. Both the hot leg and the common of a 120-volt supply must be run to the electronic thermostat.

    b. Typically there is no difference, they are designed to be interchangeable.

    c. Wire size used on the electronic thermostat must be larger.

    d. Both the low voltage hot and common must be run to the electronic thermostat.

14. A secondary drain pan is used when an air conditioning system is located:

    a. Above a ceiling

    b. In a closet

    c. Below grade

    d. More than 30 feet from a useable drain

15. A person is sitting near a cold window on a winter day. They are losing heat to the cold window by:

    a. Conduction

    b. Convection

    c. Radiation

    d. Evaporation

16. In electrical wiring, the black conductor usually indicates:

    a. The hot conductor

    b. The neutral conductor

    c. The ground wire

    d. The safety wire

Notes

17. Which of the following is the correct equation for determining wattage in an electrical circuit?

    a. $W = E \times I$

    b. $W = E \div I$

    c. $W = I \div R$

    d. $W = E \times R$

18. The rpm of a motor is determined by:

    a. The motor's voltage

    b. The motor's amperage

    c. The number of poles

    d. The size of the motor windings

19. An ohmmeter reading of infinity across a switch means:

    a. The switch is open

    b. The switch is closed

    c. The switch is good

    d. The switch is bad

20. The synchronous speed of a four-pole 60 Hz single-phase motor is:

    a. 3600 rpm

    b. 1500 rpm

    c. 3450 rpm

    d. 1800 rpm

21. Contactors are rated according to the maximum amperage through the:

    a. Coil

    b. Contacts

    c. Thermostat

    d. Fuse

22. Lowering the airflow over an evaporator will:

    a. Increase the temperature drop of the air and increase the amount of moisture removed from the air

    b. Decrease the temperature drop of the air and increase the amount of moisture removed from the air

    c. Increase the temperature drop of the air and decrease the amount of moisture removed from the air

    d. Decrease the temperature drop of the air and decrease the amount of moisture removed from the air

23. What is the temperature rise of 250 cfm of air if an electric heater adds 8,500 btuh to the air?

    a. 18 degrees F

    b. 27 degrees F

    c. 31.5 degrees F

    d. 40 degrees F

24. When a compressor service valve is located on the compressor discharge and is back-seated, _____.

    a. the compressor is connected to the discharge piping

    b. the back-seat port is closed to the compressor

    c. the front-seat port is open to the compressor

    d. All of the above are true

25. The maximum secondary amperage of a 40VA, 240 by 24 volt control transformer would be:

    a. 1.6 amps

    b. 6 amps

    c. 10.5 amps

    d. .16 amps

Notes

26. A combination electrode/nozzle gauge can be used to perform which of the following tasks?

    a. Remove the spark electrode

    b. Remove the burner nozzle head

    c. Measure the nozzle pressure

    d. Position the nozzle to the burner head

27. A single-trunk duct system that extends in one or two directions from the unit and has multiple branches is a (an):

    a. Perimeter duct system

    b. Extended plenum duct system

    c. Radial duct system

    d. Reducing trunk system

28. After cutting copper tubing, the tube edge should be cleaned with a:

    a. Wire brush

    b. Hacksaw blade

    c. Round file

    d. Reamer

29. Which of the following is covered by the National Fire Protection Code?

    a. Wire sizing

    b. Use of return air sensors

    c. Locations of service platforms

    d. Clearances around condensing units

30. Maintaining manufacturer-specified clearances on all sides of a condensing unit is:

    a. Necessary in residential applications only

    b. Necessary in commercial applications only

    c. Only necessary if required by local codes

    d. Necessary whenever locating the equipment

31. Of the following, which is most likely powered by a supply voltage of 240V single-phase?

    a. Natural gas furnace

    b. Residential condensing unit

    c. Commercial packaged unit

    d. Residential humidifier

32. Which of the following would be considered a good location to mount a wall thermostat?

    a. In a kitchen

    b. In a bathroom

    c. On an exterior wall

    d. Near a return grille

33. To have a system automatically switch to heating or cooling as the load changes, the thermostat must have:

    a. System switch set in the cooling position

    b. System switch set in the heating position

    c. Fan switch set in the auto position

    d. System switch set in the auto position

34. A heat pump operating in the heating mode with inadequate airflow across the indoor coil will typically have which of the following symptoms?

    a. A higher than normal temperature rise across the outdoor coil

    b. A lower than normal temperature rise across the outdoor coil

    c. A higher than normal temperature rise across the indoor coil

    d. A lower than normal temperature rise across the indoor coil

35. Condensate piping must be properly trapped to prevent:

    a. Ice formation in colder temperatures

    b. Condensate running off the evaporator coil

    c. Condensate overflowing the drain pan

    d. Excessive air noise in the piping

36. A BTU is defined as the amount of heat that must be added to one _____ of water to raise its temperature one degree Fahrenheit.

    a. pound

    b. ounce

    c. quart

    d. gallon

37. In a simple resistive circuit, the voltage is 24 volts and the current is 8 amps. According to Ohm's law the resistance is:

    a. 16 ohms

    b. 3 ohms

    c. 192 ohms

    d. .5 ohms

38. A traverse joint is used to connect:

    a. Electrical wiring

    b. Ductwork

    c. Refrigerant piping

    d. Condensate pipe

39. To determine the amount of sensible heat added to the refrigerant above the saturation point in an evaporator, the installer would:

    a. Perform a subcooling check on the liquid line

    b. Perform a superheat check on the evaporator line

    c. Perform a cfm check across the evaporator coil

    d. Perform a delta T check across the evaporator coil

40. To check a de-energized, normally-open relay for fused contacts, a VOM would be set to check:

    a. Amps

    b. Voltage

    c. Current

    d. Ohms

41. Which of the following oil furnace components should be replaced annually?

    a. Burner head

    b. Fuel filter

    c. Flame prover

    d. Combustion chamber

42. Pigtail 120 volt wires should be connected to the supply wiring by twisting the appropriate wires together, insulating and securing each connection with:

    a. Duct tape

    b. Plastic tape

    c. Wire nuts

    d. Rubber electrical tape

43. Which of the following components is typically used only on induced draft condensing gas furnaces?

    a. Electronic fan control

    b. Hot surface ignition

    c. Induced draft blower motor

    d. Secondary heat exchanger

44. Proper control of temperature, humidity, air movement, air cleanliness and fresh air ventilation is:

    a. To achieve a total comfort level in residential HVAC applications only

    b. To achieve a total comfort level in commercial HVAC applications only

    c. To achieve a total comfort level in any structure

    d. To achieve a total comfort level in Southern states only

45. 7,000 BTU is approximately equal to _____ KW.

    a. one

    b. two

    c. three

    d. four

46. Which precaution should an installer perform to keep copper from oxidizing internally while brazing a connector to the tube?

    a. Hold the torch flame centered over the connector

    b. Purge the tube with an inert gas such as nitrogen

    c. Increase torch flame to its maximum setting

    d. Hold the torch flame on the tubing only

47. After connecting two sections of metal ductwork using drive cleats, you should:

    a. Bend the ends of the drives over the duct

    b. Bend the ends of the drives away from the duct

    c. Cut the ends of the drives off

    d. Crimp the ends of the drives and cut off

48. The term "sheathing" in the building trade refers to:

    a. Lumber used between roof beams or rafters as a base for roofing material

    b. Lumber used to brace interior and exterior walls

    c. Interior sheet rock

    d. Plywood decking in an attic

49. Which of the following areas of metal air duct design is often governed by municipal codes?

    a. The size of air ducts

    b. The number of duct fittings

    c. Thickness of duct insulation

    d. Type of connections made at duct joints

50. A customer needs the outdoor condensing unit replaced. Which of the following should the size of the outdoor unit be based upon?

   a. The cooling load of the space

   b. The capacity of the evaporator

   c. The size of the slab

   d. The available clearances around the unit

51. Customer complaints of "cold air blowing" from the registers of a properly charged heat pump system may be rectified by:

   a. All supply registers are slightly undersized to increase air velocity

   b. Any sidewall mounted supply registers are slightly oversized to increase air velocity

   c. Only ceiling supply registers are slightly undersized to increase air velocity

   d. All supply registers are sized correctly for both throw and velocity

52. Fresh air intake ducts, used to route outdoor air to interior spaces, connect to the:

   a. Vent piping

   b. Combustion air duct

   c. Supply side of the duct system

   d. Return side of the duct system

53. According to the Uniform Mechanical Code, an attic-mounted furnace must have a permanently mounted light fixture. Where must the switch which controls this light be located?

   a. At the main circuit panel

   b. Within reach of the furnace

   c. Within sight of the furnace

   d. At the required passageway opening

54. In the following sentence, which of the choices below would best simplify and still express the nearest meaning of the statement: "In my opinion, the new thermostat design is a change for the better."

    a. An improvement

    b. A step in the right direction

    c. A help

    d. A badly needed modification

55. Metal duct that has hung for several years must be checked for sags and misalignments because:

    a. They can place excessive stress on joints and seals

    b. They increase static pressure in the duct

    c. They decrease static pressure in the duct

    d. They increase CFM at the diffusers

56. The typical procedure for repairing a small tear in the outer foil covering of a ductboard plenum is to:

    a. Cover the tear with a vapor barrier such as a piece of plastic

    b. Cover the tear with an approved foil tape

    c. Wrap the section of ductboard with insulation

    d. Cut out an area 2 inches larger than the cut and insert a replacement plug

57. A thermistor is a component used in an electronic wall thermostat to measure:

    a. Temperature

    b. Humidity

    c. Atmospheric pressure

    d. Barometric pressure

Notes

58. Two-stage heat, single-stage cool thermostats are typically used on dual fuel heat pump systems that use oil heat for second stage heating. How would the heat pump operate if the second stage of the thermostat had an open circuit on an extremely cold day? Assume the emergency heat switch is not turned on.

    a. The heat pump will operate on calls for heat without energizing the oil furnace.

    b. The oil furnace will operate on calls for heat without energizing the heat pump.

    c. Neither the heat pump nor the oil furnace will operate in heating modes.

    d. The oil furnace will operate in emergency heat mode only.

59. The slope of the condensate lines in an attic are often set by:

    a. Uniform Mechanical Code

    b. Underwriter

    c. AGA

    d. ARI

60. Which of the following joints, when properly done, will best join a copper tube to a copper fitting?

    a. Compression

    b. Flare

    c. Sweat

    d. Interference

61. Suction line refrigerant pressure and temperature are measurements needed to determine the:

    a. Air temperatures entering and leaving the evaporator coil

    b. Amount of refrigerant in a system

    c. Subcooling of refrigerant leaving a condensing coil

    d. Superheat in the refrigerant leaving an evaporator

62. On startup, a natural gas heating system has "cold air" blowing from the registers. What could be the problem?

    a. Burner manifold pressure is too high

    b. Supply duct runs are too short

    c. Blower "ON" delay is set too short

    d. Blower "ON" delay is set too long

63. A PSC condenser fan motor is not operating but the compressor runs. A voltage check of the condenser fan motor shows no voltage at the motor. Which of the following may be the cause?

    a. Open run capacitor

    b. Shorted run capacitor

    c. Open motor windings

    d. Poor or open wiring connection

64. Which of the following statements is true concerning state and local codes and regulations?

    a. Local codes never override state codes

    b. Local codes always override state codes

    c. State codes always contain regulations that are more strict than local codes

    d. Local codes must be followed if they are more strict than state codes

65. Which of the following instruments would typically be used to measure static and total pressure in a duct system?

    a. Psychrometer

    b. Diaphragm type differential pressure gauge

    c. Portable air hood

    d. Anemometer

Notes

Notes

66. Which of the following instruments must be used with the power de-energized?

    a. Voltmeter

    b. Ohmmeter

    c. Ammeter

    d. Wattmeter

67. A customer asks you to do something that your company rules specifically prohibit. When responding to their request you should:

    a. Tell the customer you cannot comply with his request and leave it at that

    b. Explore with the customer the circumstances under which you might do it

    c. Tell the customer you cannot comply with his request and offer a brief explanation why you can't

    d. Whenever possible prevent a conflict and go ahead and do it

68. Blower belt noise is caused by which of the following?

    a. Over voltage to the motor

    b. Worn pulley

    c. Under voltage to the motor

    d. Wrong blower wheel

69. While checking the installation of attic-mounted flexible duct, you find 7 to 9 foot spacing between duct supports. What steps if any, should be taken?

    a. No steps are necessary unless the spacing is greater than 10 feet

    b. No steps are necessary since the spacings are greater than 5 feet

    c. All supports must be relocated to maintain a maximum spacing of 5 feet

    d. All supports must be relocated to maintain a minimum spacing of 7 feet

70. A vapor-charged temperature control uses what in its power element?

    a. Oil

    b. Water

    c. Antifreeze

    d. Refrigerant

71. Isolation collars are typically added to ductwork to:

    a. Reduce vibration noise

    b. Reduce static pressure

    c. Reduce heat gain

    d. Increase air velocity

72. On a typical 120 volt three-prong electrical receptacle, the hot wire should be:

    a. The narrow slot

    b. The wide slot

    c. Both the narrow and wide slots

    d. The round slot

73. Each phase of three-phase power is _____ degrees out of phase from each other.

    a. 60 degrees

    b. 45 degrees

    c. 90 degrees

    d. 120 degrees

74. Two 5mfd capacitors are in series with one another. The total capacitance of the two is:

    a. 2.5mfd

    b. 5mfd

    c. 10mfd

    d. None of these

75. Three 10mfd capacitors are connected in parallel with each other. The total capacitance of the three is:

    a. 10mfd

    b. 20mfd

    c. 30mfd

    d. None of these

76. The air temperature drop across an evaporator is measured and determined to be 30 degrees Fahrenheit. Which of the following is most likely true?

    a. There is too much evaporator airflow

    b. There is too little evaporator airflow

    c. The system is overcharged

    d. The system is undercharged

77. The square root of the number 9 is:

    a. 1.5

    b. 2

    c. 3

    d. 6

78. Air flows from an 8-inch duct where the duct is reduced to 6 inches. Which of the following is true?

    a. The air velocity increases

    b. The air velocity decreases

    c. The CFM increases

    d. The CFM decreases

79. Which of the following is true about compression ratio?

    a. It is the absolute high side pressure divided into the absolute suction pressure.

    b. It is the absolute suction pressure divided into the absolute high side pressure.

    c. It is the high side pressure in psig divided into the suction pressure in psig.

    d. It is the suction pressure in psig divided into the high side pressure in psig.

80. The speed of a three-phase motor is determined by:

    a. The voltage to the motor

    b. The current draw of the motor

    c. The frequency of the power supplied to the motor

    d. The number of poles in the motor

81. Which one of the following fuse types should be used to protect the condensing unit on a residential split system air conditioning system?

    a. Time-delay

    b. Instant-blow

    c. Fast-reacting

    d. Series-rated ground-fault

Notes

82. Which of the following organizations rates electrical components and equipment for electrical safety?

    a. ARI

    b. UL

    c. RSES

    d. ASHRAE

83. On a heat pump the "balance point" is:

    a. The point at which the heat loss of a structure matches the heat pump capacity

    b. The point at which the heating load calculation matches the cooling load calculation

    c. The lowest effective outside temperature at which the heat pump can operate

    d. The most effective temperature where the heat pump's electric auxiliary heat should be energized

84. On a system using a reciprocating compressor, raising the evaporator pressure and/or lowering the condensing pressure will:

    a. Decrease the compressor's volumetric efficiency

    b. Increase the compressor's volumetric efficiency

    c. Have no effect of volumetric efficiency

    d. Increase the compressor's power consumption

85. A heat anticipator in a typical thermostat is wired:

    a. In series with the "R" terminal

    b. In parallel with the "R" terminal

    c. In series with the "W" terminal

    d. In parallel with the "W" terminal

86. To test a capacitor with an ohmmeter, the ohmmeter should be set at:

    a. The R times 1 scale

    b. The R times 10 scale

    c. The R times 100 scale

    d. The highest ohms scale available

87. Which of the following is considered a good evacuation?

    a. 2000 microns

    b. 1500 microns

    c. 1000 microns

    d. 500 microns or less

88. A filter-drier has a temperature drop across it. Which of the following is true about the filter-drier?

    a. It is partially plugged

    b. It is totally plugged

    c. A temperature drop across it is normal

    d. The filter-drier has not yet broken in

89. An FRN-R fuse is:

    a. A time-delay fuse

    b. An instant-blow fuse

    c. Neither A nor B is true

    d. Only used on non-inductive circuits

Notes

90. A fuse must be selected according to:

    a. Type, amperage and voltage

    b. Type and amperage

    c. Type and voltage

    d. Type, amperage, voltage and wattage

91. On a typical three-terminal single-phase fully hermetic compressor, between which two terminals is the start winding located?

    a. The two with the highest resistance

    b. The two with the least resistance

    c. The two with the middle resistance

    d. The two with continuity to ground

92. A Ground Fault Circuit Interrupter monitors:

    a. The current between the hot wire and ground

    b. The current between the hot wire and the neutral wire

    c. The current between the neutral wire and ground

    d. The voltage across the line

93. Typical bypass humidifiers are connected:

    a. Between the supply and the return ducts

    b. Between the supply and outside air intake

    c. Between the outside air intake and return air duct

    d. Between the exhaust air and return air ducts

94. The typical bypass humidifier does not work well with which of the following heating system types?

    a. Natural gas

    b. Propane gas

    c. Heat pumps and electric heat

    d. Oil-fired furnaces

95. During the heating season a damper located in the bypass duct of a bypass humidifier must be:

    a. Closed

    b. Open

    c. Adjusted according to the temperature rise of the furnace

    d. Adjusted according to the total cfm of the duct system.

96. As air is heated, its relative humidity:

    a. Increases

    b. Decreases

    c. Remains the same

    d. None of these is correct

97. A sling Psychrometer measures:

    a. Dew point temperature

    b. Specific humidity

    c. Relative humidity

    d. Dry and wet bulb temperatures

Notes

98. A protractor is best used to:

    a. Measure distance

    b. Lay out angles

    c. Determine air velocity

    d. Select a contractor

99. Which of the following is the most accurate method of balancing a duct system?

    a. Proportional balancing

    b. Sequential balancing

    c. Temperature balancing

    d. Total pressure balancing

100. Which of the following statements most correctly defines the difference between "accuracy" and "precision" in instruments?

    a. There is no difference between the two terms.

    b. "Accuracy" refers to how close the measurement is to the actual value being measured while "precision" refers to how small a value can be measured.

    c. "Accuracy" refers to how small a value can be measured while "precision" refers to how close the reading is to the actual value.

    d. "Accuracy" refers to the repeatability of the measurement while "precision" refers to the percentage of error in the reading.

PART THIRTEEN
Answer Keys to the Practice Exams

PART THREE
Answer Key to the Core Practice Exam

1. a

2. c

3. c

4. b

5. b

6. b

7. b

8. a

9. a

10. c

11. b

12. a

13. c

14. a

15. c

16. b

17. c

18. c

19. b

20. a

21. c

22. c

23. a

24. b

25. d

Notes

PART THREE
Answer Key to the Core Practice Exam (*cont.*)

26. b

27. c

28. a

29. a

30. a

31. a

32. a

33. a

34. b

35. a

36. a

37. b

38. b

39. b

40. b

41. b

42. a

43. a

44. b

45. a

46. a

47. a

48. a

49. c

50. d

Notes

PART FOUR
Answer Key to the Math Practice Exam

1. c

2. a

3. d

4. b

5. c

6. d

7. c

8. a

9. a

10. b

11. b

12. c

13. c

14. a

15. b

16. d

17. c

18. c

19. b

20. d

21. a

22. d

23. d

24. c

25. d

Notes

PART FOUR
Answer Key to the Math Practice Exam (*cont.*)

26. b

27. c

28. d

29. c

30. a

31. a

32. d

33. b

34. b

35. c

36. d

37. a

38. c

39. a

40. b

41. d

42. b

43. b

44. b

45. b

46. d

47. c

48. a

49. d

50. b

Notes

PART FIVE
Answer Key to the Electrical Practice Exam

1. c

2. a

3. b

4. a

5. a

6. d

7. b

8. a

9. b

10. b

11. d

12. a   *Direct acting means "closes on rise". Contacts close on an increase.*

13. d

14. b

15. a

16. a

17. a

18. c

19. a

20. d

21. d

22. b

23. b

24. d

25. b

26. c

27. b   *There are 3.41 btu per watt and 1000 watts per kilowatt*

28. b

29. d

30. c

31. c

Notes

PART FIVE
Answer Key to the Electrical
Practice Exam (*cont.*)

32. b

33. b

34. a

35. b

36. b *Reverse acting means "opens on rise". Opens on an increase in pressure.*

37. a *Back to back scrs allow alternating current to pass but can be controlled by the switching of the gates to allow for high speed switching.*

38. b

39. d *If you have trouble with this question you need to study the fan laws and work out a few fan law problems.*

40. b

41. b

42. c

43. b

44. b

45. c *When the internal compressor thermostats or overloads open they shut down the compressor but the control circuit remains in the circuit. There is no oil pressure with the compressor off so the oil failure control trips and locks out the control circuit. By the time the technician arrives the internal thermostats in the compressor have cooled and reset. The technician just finds the compressor out on the oil failure control and can be led to believe that an oil problem exists when the actual problem is compressor overheating.*

46. a

47. b

48. c

49. b

50. a

Notes

PART SIX
Answer Key to the Air Conditioning Practice Exam #1

1. a
2. a
3. b
4. a
5. d
6. a
7. d
8. a
9. d
10. a
11. a
12. d
13. b
14. d
15. c
16. a
17. b
18. b
19. a
20. a
21. b
22. a
23. b
24. b
25. a   *The anemometer only measures feet while the operator uses a stop watch or second hand on a watch. This is a typical "trick" question to watch for.*

Notes

PART SIX
Answer Key to the Air Conditioning Practice Exam #1 (*cont.*)

26. c

27. b

28. c

29. b

30. b

31. c

32. d

33. a

34. b

35. d

36. a

37. a

38. a

39. a

40. d

41. d

42. d

43. b

44. c

45. a

46. a

47. c

48. c

49. c

50. d

Notes

PART SIX
Answer Key to the Air Conditioning Practice Exam #2

1. b

2. c

3. b

4. a

5. b

6. b

7. c

8. b

9. c

10. c

11. a

12. c

13. c

14. d

15. d

16. b

17. d

18. c

19. a

20. a

21. a

22. d

23. c

24. b

25. b

Notes

PART SIX
Answer Key to the Air Conditioning Practice Exam #2 (*cont.*)

26. b

27. b

28. a

29. c

30. d

31. b

32. c

33. c

34. a

35. b

36. d

37. b

38. a

39. a

40. d

41. d

42. a

43. c

44. a

45. a

46. c

47. c

48. a

49. d

50. c

Notes

PART SEVEN
Answer Key to the Heat Pump Practice Exam

1. b

2. a

3. c

4. d

5. a

6. d

7. d

8. d

9. d

10. b

11. d

12. b

13. a

14. d

15. a

16. d

17. a

18. b

19. a

20. b

21. a

22. b

23. c

24. b

25. b

Notes

PART SEVEN
Answer Key to the Heat Pump
Practice Exam (*cont.*)

26. d

27. a

28. d

29. c

30. a

31. a

32. d

33. a

34. a

35. a

36. b

37. b

38. a

39. b

40. b

41. c

42. c

43. a

44. d

45. c

46. c

47. d

48. c

49. b

50. a

Notes

PART EIGHT
Answer Key to the Gas Heating Practice Exam

1. c

2. c

3. b

4. d

5. c

6. c

7. c

8. b

9. b

10. b

11. b

12. a

13. a

14. a

15. b

16. b

17. d

18. b

19. a

20. b

21. a

22. d

23. b

24. a

25. d

Notes

PART EIGHT
Answer Key to the Gas Heating
Practice Exam (*cont*.)

26. a

27. a

28. d

29. c

30. d

31. a

32. b

33. a

34. c

35. d

36. d

37. a

38. a

39. c

40. b

41. c

42. a

43. c

44. a

45. b

46. c

47. b

48. b

49. a

50. d

Notes

PART NINE
Answer Key to the Air Distribution Practice Exam

1. a

2. b

3. a

4. b

5. a

6. a

7. a

8. b

9. a

10. b

11. b

12. b

13. a

14. b

15. b

16. d

17. a

18. a

19. d

20. d

21. b

22. a

23. c

24. c

25. a

Notes

PART NINE
Answer Key to the Air Distribution Practice Exam (*cont.*)

26. b

27. d

28. a

29. c

30. b

31. c

32. b

33. d

34. a

35. d

36. b

37. b

38. b

39. b

40. b

41. a

42. c

43. b

44. c

45. d

46. d

47. c

48. c

49. a

50. b

Notes

PART TEN
Answer Key to the Controls

1. a

2. a

3. b

4. a

5. b

6. b

7. b

8. a

9. c

10. b

11. a

12. a

13. b

14. a

15. b

16. a

17. a

18. a

19. b

20. a

21. b

22. b

23. a

24. b

25. d

Notes

PART TEN
Answer Key to the Controls (*cont.*)

26. a

27. b

28. a

29. a

30. a

31. a

32. a

33. a

34. a

35. b

36. a

37. b

38. b

39. c

40. a

41. a

42. a

43. d

44. a

45. a

46. b

47. a

48. a

49. b

50. a

Notes

PART ELEVEN
Answer Key to the Indoor Air Quality

1. c

2. a

3. a

4. d

5. c

6. a

7. b

8. d

9. a

10. d

11. d

12. d

13. c

14. a

15. a

16. b

17. a

18. b

19. c

20. a

21. c

22. d

23. a

24. a

25. b

Notes

PART ELEVEN
Answer Key to the Indoor Air Quality (*cont*.)

26. a

27. d

28. a

29. a

30. c

31. b

32. d

33. b

34. d

35. a

36. d

37. d

38. d

39. c

40. c

41. a

42. a

43. a

44. a

45. a

46. a

47. a

48. a

49. a

50. a

Notes

PART TWELVE
Answer Key to the Short Final Practice Exam

1. d

2. b

3. c

4. a

5. d

6. c

7. b

8. d

9. a

10. d

11. c

12. b

13. b

14. c

15. c

16. a

17. a

18. d

19. c

20. b

21. d

22. c

23. d

24. c

25. a

Notes

PART TWELVE
Answer Key to the Short Final Practice Exam (*cont.*)

26. c

27. b

28. a

29. b

30. c

31. b

32. a

33. d

34. a

35. a

36. c

37. a

38. d

39. d

40. b

41. d

42. a

43. c

44. a

45. d

46. c

47. b

48. c

49. d

50. a

PART TWELVE
Answer Key to the Full-Length Final Practice Exam

1. a

2. a

3. c

4. a

5. c

6. a

7. b

8. c

9. a

10. a

11. b

12. c

13. b

14. a

15. c

16. a

17. a

18. c

19. a

20. d

21. b

22. a

23. c

24. d

25. a

Notes

PART TWELVE
Answer Key to the Full-Length Final Practice Exam (*cont.*)

26. d

27. b

28. d

29. b

30. d

31. b

32. d

33. d

34. c

35. c

36. a

37. b

38. b

39. b

40. d

41. b

42. c

43. c

44. c

45. b

46. b

47. a

48. a

49. c

50. b

Notes

**PART TWELVE
Answer Key to the Full-Length
Final Practice Exam (*cont.*)**

51. d

52. d

53. d

54. a

55. a

56. b

57. a

58. a

59. a

60. c

61. d

62. c

63. d

64. d

65. b

66. b

67. c

68. b

69. d

70. d

71. a

72. a

73. d

74. a

75. c

Notes

PART TWELVE
Answer Key to the Full-Length
Final Practice Exam (*cont.*)

76. b

77. c

78. a

79. b

80. c

81. a

82. b

83. a

84. b

85. c

86. d

87. d

88. a

89. a

90. a

91. c

92. b

93. a

94. c

95. b

96. b

97. d

98. b

99. a

100. b

Notes

PART FOURTEEN
Equations

- HVAC Electrical Equations

- HVAC Fan Laws

- HVAC Air Side Equations

- Sheet Metal Equations

- HVAC Mechanical Equations

- HVAC Hydronics Equations

- Heat Transfer Equations

HVAC ELECTRICAL EQUATIONS

$1\phi BHP = \dfrac{E \times I \times \%EFF \times PF}{746}$	$3\phi BHP = \dfrac{E \times I \times \%EFF \times PF \times 1.732}{746}$
$E = IR$	$R = \dfrac{E}{I}$
$I = \dfrac{E}{R}$	$BHP_2 = Name\ Plate\ HP \times \dfrac{Actual\ Amps}{FLA}$
$BHP_2 = BHP_1 \times \left(\dfrac{RPM_2}{RPM_1}\right)^3$	$Approx\ BHP = \dfrac{E \times I}{746}$
$I_2 = I_1 \times \left(\dfrac{CFM_2}{CFM_1}\right)^3$	$I_2 = I_1 \times \left(\dfrac{RPM_2}{RPM_1}\right)^3$
$3\phi\%E_{Imbal} = \dfrac{Max\ E_{Diff}}{Ave\ E} \times 100$	$3\phi\%I_{Imbal} = \dfrac{Max\ I_{Diff}}{Ave\ I} \times 100$
$\%Temp\ Rise = 2 \times \left(\%E_{Imbal}\right)^2$	$1\phi PF = \dfrac{W}{VA}$
$3\phi PF = \dfrac{W}{VA \times 1.732}$	$R_T = R_1 + R_2 + R_3 \dots$
$R_T = \dfrac{R_1 \times R_2}{R_1 + R_2}$	$R_T = \dfrac{1}{\frac{1}{R_1} + \frac{1}{R_2} + \frac{1}{R_3} \dots}$

HVAC ELECTRICAL EQUATIONS (cont.)

$$C_T = C_1 + C_2 + C_3 \ldots$$	$$C_T = \frac{C_1 \times C_2}{C_1 + C_2}$$
$$C_T = \frac{1}{\frac{1}{C_1} + \frac{1}{C_2} + \frac{1}{C_3} \ldots}$$	$$MFD = \frac{2650 \times I}{E}$$
$$X_L = 2\pi FL$$	$$Z = \sqrt{R^2 + \left(X_L - X_C\right)^2}$$
$$RPM = 120 \times \left(\frac{Freq}{Poles}\right)$$	$$X_C = \frac{1}{2\pi FC}$$
$$Z = \sqrt{R^2 + X_{L^2}}$$	$$1\phi VD = \frac{2 \times R \times I \times L}{CM}$$
$$\frac{N_P}{N_S} = \frac{E_P}{E_S}$$	$$Z = \sqrt{R^2 + X_{C^2}}$$
$$3\phi VD = \frac{1.732 \times R \times I \times L}{CM}$$	$$\frac{E_P}{E_S} = \frac{I_S}{I_P}$$
$$\%EFF = \frac{Energy\ Out}{Energy\ In}$$	$$VD = I^2 R$$

HVAC FAN LAWS

$$\frac{CFM_1}{CFM_2} = \frac{RPM_1}{RPM_2}$$	$$\frac{DIA_A}{DIA_B} = \frac{RPM_B}{DIA_A}$$
$$\left(\frac{CFM_1}{CFM_2}\right)^2 = \frac{SP_1}{SP_2}$$	$$\frac{CFM_1}{CFM_2} = \sqrt{\frac{SP_1}{SP_2}}$$
$$\left(\frac{CFM_1}{CFM_2}\right)^3 = \frac{BHP_1}{BHP_2}$$	$$\frac{CFM_1}{CFM_2} = \sqrt[3]{\frac{BHP_1}{BHP_2}}$$
$$CFM_2 = CFM_1 \times \frac{RPM_2}{RPM_1}$$	$$CFM_2 = CFM_1 \times \sqrt{\frac{SP_2}{SP_1}}$$
$$CFM_2 = CFM_1 \times \sqrt[3]{\frac{BHP_2}{BHP_1}}$$	$$BHP_2 = BHP_1 \times \left(\frac{CFM_2}{CFM_1}\right)^3$$
$$BHP_2 = BHP_1 \times \left(\frac{RPM_2}{RPM_1}\right)^3$$	$$BHP_2 = BHP_1 \times \left(\frac{D_2}{D_1}\right)$$
$$RPM_2 = RPM_1 \times \left(\frac{CFM_2}{CFM_1}\right)$$	$$SP_2 = SP_1 \times \left(\frac{CFM_2}{CFM_1}\right)^2$$
$$Max\ Motor\ Sheave\ Dia = Exist\ Dia \times \sqrt[3]{\frac{Max\ BHP}{Ex\ BHP}}$$	

HVAC AIR SIDE EQUATIONS

$$CFM = \frac{BTUH}{1.08 \times \Delta T}$$	$$CFM = \frac{E \times I \times 3.414}{1.08 \times \Delta T}$$
$$FPM = 4005 \times \sqrt{V_P}$$	$$FPM = 1096.7 \times \sqrt{\frac{V_P}{DEN}}$$
$$CFM = CuFt \times \frac{AC/HR}{60}$$	$$V_P = \left(\frac{FPM}{4005}\right)^2$$
$$CFM = A_{Sq\,Ft} \times V_{FPM}$$	$$A_{Sq\,Ft} = \frac{CFM}{FPM}$$
$$V_{FPM} = \frac{CFM}{A}$$	$$\%OA = \frac{RAT - MAT}{RAT - OAT} \times 100$$
$$MAT = OAT - \left[(OAT - RAT) \times \%RA\right]$$	$$MAT = \frac{(\%OA \times OAT) + (\%RA \times RAT)}{100}$$

SHEET METAL EQUATIONS

$Sq\ Inches = Sq\ Ft \times 144$	$A_{Sq\ Ft} = \dfrac{H'' \times W''}{144}$
$Sq\ Feet = \dfrac{Sq\ Inches}{144}$	$Round\ Area = 3.1416 \times D^2$
$Radius = \sqrt{\dfrac{Area}{3.1416}}$	$Round\ Area\ Sq\ Ft = \dfrac{3.1416 \times R^2}{144}$
$Radius = \dfrac{C}{6.28}$	$Aspect\ Ratio = \dfrac{Width}{Height}$
$W = \dfrac{Area}{Height}$	$FPM = \dfrac{CFM}{Area}$
$CFM = FPM \times Area$	$D = 2 \times R$
$Area = \dfrac{CFM}{FPM}$	$D'' = \sqrt{\dfrac{4 \times H \times W}{3.1416}}$
$C'' = 3.1416 \times D''$	$Vol = L \times W \times H$
$Vol = 3.1416 \times R^2 \times H$	$R = \sqrt{\dfrac{Vol}{3.1416 \times H}}$
$C^2 = A^2 + B^2$	$C = \sqrt{A^2 + B^2}$

HVAC MECHANICAL EQUATIONS

$Belt\ Length = 2c + \left[1.57 \times (D+d)\right] + \dfrac{(D-d)^2}{4c}$	$Arc\ of\ Contact = 180 - \left[\dfrac{(D-d) \times 60}{C}\right]$
$Belt\ Length = Existing\ Length \times 1.57 \times (Dia_1 - Dia_2)$	$Belts = \dfrac{HP}{(HP \div Belt) \times SF}$
$Compression\ Ratio = \dfrac{Head\ Pressure\ PSIA}{Suction\ Presssure\ PSIA}$	$PSIA = PSIG + 14.7$
$PSIA = \dfrac{30 - Inches\ HG}{2}$	$\dfrac{HP}{Ton} = \dfrac{4.71}{Cop}$
$COP = \dfrac{BTUH\ Output}{Input\ Watts \times (3.414\ BTU \div W)}$	$COP = (SEER \times 2.93) - 1$
$SEER = \dfrac{COP + 1}{.293}$	$EER = \dfrac{BTUH\ Output}{Input\ Watts}$
$HSPF = \dfrac{BTUH\ Output}{BTUH\ Input}$	$AFUE = \dfrac{Annual\ Energy\ Output}{Annual\ Energy\ Input}$

HVAC HYDRONICS EQUATIONS

$$BTUH = 500 \times GPM \times \Delta T$$	$$GPM = \frac{BTUH}{500 \times \Delta T}$$
$$\Delta T = \frac{BTUH}{GPM \times 500}$$	$$GPM = C_V \times \sqrt{\Delta P}$$
$$C_V = \frac{GPM}{\sqrt{\Delta P}}$$	$$GPM_2 = GPM_1 \times \sqrt[3]{\frac{BHP_2}{BHP_1}}$$
$$BHP = \frac{GPM \times TDH}{3960 \times EFF}$$	$$\frac{BHP_2}{BHP_1} = \left(\frac{RPM_2}{RPM_1}\right)^3$$
$$\frac{BHP_2}{BHP_1} = \left(\frac{GPM_2}{GPM_1}\right)^3$$	$$\frac{BHP_2}{BHP_1} = \left(\frac{D_2}{D_1}\right)^3$$
$$\frac{GPM_2}{GPM_1} = \frac{RPM_2}{RPM_1}$$	$$\frac{GPM_2}{GPM_1} = \frac{D_2}{D_1}$$
$$\frac{\Delta P_2}{\Delta P_1} = \left(\frac{GPM_2}{GPM_1}\right)^2$$	$$\frac{H_2}{H_1} = \left(\frac{RPM_2}{RPM_1}\right)^2$$
$$\frac{\Delta P_2}{\Delta P_1} = \left(\frac{RPM_2}{RPM_1}\right)^2$$	$$\frac{H_2}{H_1} = \left(\frac{GPM_2}{GPM_1}\right)^2$$
$$MWT = \frac{(GPM_1 \times T_1) + (GPM_2 \times T_2)}{GPM\ Total}$$	

HEAT TRANSFER EQUATIONS

$BTU = SPHT \times Lbs \times \Delta T$	$BTUH = 1.08 \times CFM \times \Delta T$
$SPHT = \dfrac{BTU}{Lbs \times \Delta T}$	$\Delta T = \dfrac{BTUH}{1.08 \times CFM}$
$BTUH = 4.5 \times CFM \times \Delta H$	$CFM = \dfrac{BTUH}{1.08 \times \Delta T}$
$\dfrac{Lbs}{MW} = CFM \times .075$	$\dfrac{Lbs}{MW} = \dfrac{200\ BTU \div MW}{NRE\ IN\ BTU \div Lb}$
$BTUH = 500 \times GPM \times \Delta T$	$GPM = \dfrac{BTUH}{500 \times \Delta T}$
$BTUH = U \times A \times \Delta T$	$LMTD = \dfrac{\Delta T_L - \Delta T_S}{L_N \left(\Delta T_L \div \Delta T_S \right)}$
$U = \dfrac{1}{R_T}$	$R_T = \dfrac{1}{U}$
$NRE = HG - HF$	$Tons = \dfrac{GPM \times TD}{24}$
$BF = \dfrac{LADB - ECT}{EADB - ECT}$	$SHF = \dfrac{BTUH\ Sensible}{BTUH\ Latent}$
$Linear\ Expansion = Coefficient\ of\ Expansion \times L \times TD$	

Notes

PART FIFTEEN
Charts, Tables and Reference Materials

- Useful HVAC Values and Multipliers
- Common HVAC Acronyms
- Common ARI and ASHRAE Standards
- Refrigerant Classifications
- Efficiency Ratings
- 410A/22 Temperature–Pressure Chart

USEFUL HVAC VALUES AND MULTIPLIERS

3.414 BTU per Watt

One Kilowatt = 3415 Btuh

746 Watts per horsepower

One Ton of cooling equals
 288,000 Btu per 24 hours
 12,000 Btuh
 200 Btu per minute

1 Therm = 100,000 BTU

8.33 pounds per gallon of water

1 cubic foot of water = 7.48 gallons

1 cubic foot of water = 62.4 pounds (62.37 exactly)

7000 grains = 1 pound

One horsepower per Ton @40 degree evaporator saturation temperature

Atmospheric pressure @ sea level is 14.696 psia

Gauge pressure = psia minus 14.7

Psia = psig plus 14.7

One psi = 2.31 feet of water

One foot of water = .433 psi

.24 specific heat of air

.075 pounds per cubic foot (Density of standard air)*

13.33 cubic feet per pound (Specific volume of standard air)*

25,400 microns per inch

29.92 inches Hg (Standard sea level air pressure)*

8760 hours in a year

R-value of an inside air film is .68

*Standard air is air at 70 degrees Fahrenheit dry bulb, 0% relative humidity, at sea level.

COMMON HVAC ACRONYMS

OA	Outside Air		A	Area
SA	Supply Air		TP	Total Pressure
RA	Return Air		SP	Static Pressure
EA	Exhaust Air		VP	Velocity Pressure
SAT	Supply Air Temperature		PSIG	Pounds per Square inch Gauge
MAT	Mixed Air Temperature		PSIA	Pounds per Square inch Atmospheric or Absolute
OAT	Outside Air Temperature		MBH	Thousands of BTU per Hour
RAT	Return Air Temperature		WB	Wet Bulb temperature
R/E	Return and Exhaust		DB	Dry Bulb temperature
HWS	Hot Water Supply		RH	Relative Humidity
HWR	Hot Water Return		DP	Dew Point temperature
CS	Condenser Supply		KW	Kilo-Watts
CR	Condenser Return		W	Watts
NPHP	Name Plate Horsepower		I	Amperage or current
TDH	Total Dynamic Head		TD	Temperature Difference
NPSH	Net Positive Suction Head (slang = Not Pumping So Hot)		DELTA T	Temperature Difference (ΔT)
BHP	Brake Horsepower		Eff	Efficiency
FLA	Full Load Amps		Hz	Hertz or cycles per second
LRA	Locked Rotor Amps		PF	Power Factor
FPM	Feet Per Minute		SF	Service Factor
CFM	Cubic Feet per Minute		SSH	Static Suction Head
GPM	Gallons Per Minute		STH	Static Total Head
Wg	Water Gauge		SVH	Static Velocity Head
Hg	Mercury		SST	Saturated Suction Temperature
Q	Quantity		SDT	Saturated Discharge Temperature
V	Velocity			

COMMON ARI AND ASHRAE STANDARDS

ASHRAE Standard 34

The American Society of Refrigeration and Air Conditioning Engineers has classified all refrigerants as to their toxicity and flammability.

Toxicity is based upon the level to which an individual can be exposed over his lifetime without ill effects and is stated as its (TLV) Threshold Limit Value. Refrigerants are also rated according to their (TWA) Time Weighted Average.

- Class A refrigerants are those which show no evidence of toxicity at concentrations below 400 parts per million. (Most refrigerants are Class A)
- Class B refrigerants are those that show evidence of toxicity at concentrations below 400 parts per million.

Flammability falls into three groups or classes.

- Class 1 refrigerants show no flame propagation when tested in air at 65 degrees @ 14.7 psia.
- Class 2 refrigerants are flammable and require caution.
- Class 3 refrigerants are considered highly flammable and require even greater caution.

Example: A refrigerant may be classified as a B-1 refrigerant. Such a refrigerant is toxic but is not flammable. Another refrigerant may be an A-3 refrigerant which indicates it is not considered toxic but is highly flammable.

ASHRAE Standard 15

Deals with equipment room safety and requires the use of sensors and alarms to detect refrigerant leaks in equipment rooms.

- This standard applies to all refrigerants in all classifications.
- Each equipment room must have an alarm that will activate before the refrigerant concentrations exceed that particular refrigerant's TLV and TWA.
- Every system installed in an equipment room must be equipped with a safety pressure relief valve vented to the outdoors.
- Equipment rooms must be provided with ventilation meeting ASHRAE Standard 15R.
- At least one approved self-contained breathing apparatus (SCBA) should be located near an equipment room.
- The lower limit of acceptable oxygen levels in a room is 19.5%. (20.9% is normal).

ARI Standard 700

This is the EPA's accepted standard for the level of purity of refrigerants. Refrigerants must meet this standard or it cannot be sold to a customer. Only the refrigerant manufacturers have the ability (chemical labs) necessary to properly test for this purity level. Technicians must be aware that refrigerant failing to meet this standard cannot be sold. It is legal for a technician to recover, filter and recharge refrigerant into equipment owned by the same customer without meeting the ARI 700 standard.

ARI Standard 740

Refrigerant recovery equipment must be certified by an independent agency that the equipment conforms to this standard. Technicians must be aware that recovery equipment must be listed as certified to ARI Standard 740.

REFRIGERANT CLASSIFICATION ACCORDING TO FLAMMABILITY AND TOXICITY

	Low Toxicity	High Toxicity
Higher Flammability	A3	B3
Lower Flammability	A2	B2
No Flammability	A1	B1

All refrigerants are rated according to their safety as per their flammability and toxicity. Federal, state and local codes may require that refrigerants of a particular safety classification be treated according to particular regulations. For example, some codes require that systems containing B3 refrigerants have constant leakage monitors with alarms installed in equipment rooms.

Refrigerant	Classification
R-11	A1
R-12	A1
R-22	A1
R-123	B1
R-134a	A1
R-401a	A1
R-406a	A2
R-500	A1
R-502	A1
R-507	A1
R-717	B2
R-744	A1

EFFICIENCY RATINGS

What They Mean

The heating and cooling industry has used a number of efficiency ratings (each similar but slightly different). The U.S. Department of energy has mandated by law that heating and cooling equipment be tested and meet minimum efficiency standards. Consumers can compare equipment of different efficiencies by comparing equivalent ratings. A basic understanding of the ratings, how they are determined and where they apply is necessary to their use. Each rating system is explained below.

COP Coefficient Of Performance

- Is applied to mechanical cooling systems using refrigerants which also includes high, medium and low temperature systems as well as heat pumps.

- Is the ratio of how much energy a system uses versus how much energy the system produces.

- Is found by dividing the output by the input.

- For example, a coefficient of 2.5 indicates that a heat pump produces 2.5 times more heat output in Btuh than its electrical input in Btuh.

COP = Btuh Output divided by Btuh Input

OR

COP = Btuh Output divided by (Input Watts times 3.42 btu/watt)

- Ground source heat pumps have the highest Coefficient Of Performance, obtaining COPs as high as 5 as of 2004. A COP of 5 indicates that the heat pump is providing five times as much heat output as it consumes in electrical power.

- Typical air to air heat pumps have COPs ranging between 2 and 3.5.

- The COP listed on the system is not the actual operating COP. An incorrectly sized and/or installed system will operate at a lower efficiency than its rating. Proper selection and installation is critical.

- A field determined COP after the installation is complete is the only actual guarantee that the system is performing at its expected efficiency.

EER Energy Efficiency Ratio

- Is applied to mechanical cooling systems using refrigerants. This is an older efficiency rating system which has been replaced with SEER.

- Is the ratio of the cooling capacity a system provides as compared to its energy input.

- Is found by dividing the name plate output by the name plate input.

- The EER fails to consider operational factors such as geographical location, seasonal differences and cycling rates.

EER = Btuh Output divided by Input Watts

EFFICIENCY RATINGS (cont.)

SEER Seasonal Energy Efficiency Ratio

- Is applied to mechanical cooling systems using refrigerants. Also used to compare heat pump efficiency in the cooling cycle.

- The SEER rating is found by dividing the output by the input. However, in addition the SEER rating is adjusted to take into consideration seasonal operational variations and system cycling rates. Therefore, the SEER rating is more realistic and has replaced the EER rating.

SEER = Btuh Output divided by Input Watts
(adjusted to reflect operational variables)

- SEER ratings for air conditioning equipment prior to 1992 were in the range of 6 to 8 SEER. Systems manufactured between 1992 and 1998 had SEER ratings between 10 and 13 SEER. SEER ratings have risen to as high as 20 SEER as of 2004. The higher the SEER rating the more the energy savings. However, the initial purchase cost is also higher. A careful calculation comparing initial cost to energy savings can determine the most cost effective SEER for a particular application. The current minimum SEER as mandated by the DOE is 10 SEER. As of January 2006, the minimum SEER will increase to 13.

- The SEER listed on the system is not the actual operating SEER. An incorrectly sized and/or installed system will operate at a lower efficiency than its rating. Proper selection and installation is critical.

HSPF Heating Seasonal Performance Factor

- Applies to the efficiency of heat pumps.

- The HSPF is the ratio of the btuh heating output capacity of a heat pump to the energy input. The HSPF only applies to the heating cycle of the heat pump operation.

- The HSPF is an adjusted ratio taking into consideration seasonal operational variations and system cycling rates. Therefore, the HSPF is a fairly realistic energy efficiency rating.

HSPF = Btuh Output divided by Btuh Input
(adjusted to reflect operational variables)

- The HSPF listed on the heat pump is not the actual operating HSPF. An incorrectly sized and/or installed system will operate at a lower efficiency than its rating. Proper selection and installation is critical.

EFFICIENCY RATINGS (cont.)

AFUE Annual Fuel Utilization Efficiency

- Applies to fuel burning furnaces, boilers and other combustion heating equipment.

- The AFUE compares the annual output energy of the system to the annual input of energy. Since the AFUE rating is averaged for an entire year, it takes into consideration variations in operation due to starting, stopping and seasonal differences.

- The AFUE is expressed in percent. For example, a natural gas furnace with a 60% AFUE indicates that the system is only 60% efficient with the remaining 40% of unburned fuel lost up the stack.

- Since 1992 all furnaces must have an AFUE of no less than 78%.

- Furnaces with AFUE ratings exceeding 95% are readily available.

- Furnaces manufactured prior to 1985 usually have an AFUE in the 50 to 60% range. By 1990, the typical efficiency had risen to around 80%.

AFUE = Annual Energy Output divided by Annual Energy Input

- The purchase and installation of a furnace with a high AFUE is no guarantee that it will actually operate at the listed AFUE. An incorrectly sized furnace for the structure and/or a poor installation will degrade the actual operating AFUE. The AFUE listed on any furnace assumes it is correctly matched for the application and is properly installed.

- Note: Replacing a typical residential 60% AFUE gas furnace with a 95% AFUE furnace reduces greenhouse gas emissions (Carbon Dioxide) into the atmosphere by approximately 1,500 pounds per year. Higher AFUE furnaces greatly reduce pollution and help prevent global warming.

Converting SEER to COP

A Seasonal Energy Efficiency Ratio may be converted to its Coefficient Of Performance using the following equation.

COP = (SEER X .293) minus 1

Example: A heat pump has a SEER rating of 12 in the heating cycle. What is its COP?

COP = (12 X .293) minus 1
COP = 3.561 minus 1
COP = 2.561

410A/22 TEMPERATURE–PRESSURE CHART

Temperature (°F)	R-22 Pressure (psig)	R-410A Pressure (psig)	Temperature (°F)	R-22 Pressure (psig)	R-410A Pressure (psig)
-30	4.7	17.8	26	50	89.1
-28	5.9	19.4	28	52.4	92.9
-26	6.9	21.1	30	55	96.8
-24	8	22.7	32	57.5	100.9
-22	9.1	24.5	34	60.2	105
-20	10.2	26.3	36	62.9	109.2
-18	11.4	28.2	38	65.7	113.6
-16	12.6	30.1	40	68.6	118.1
-14	13.9	32.2	42	71.5	122.7
-12	15.2	34.2	44	74.5	127.4
-10	16.5	36.4	46	77.6	132.2
-8	17.9	38.6	48	80.8	137.2
-6	19.4	40.9	50	84.1	142.2
-4	20.9	43.3	55	92.6	155.5
-2	22.4	45.7	60	101.6	169.6
0	24	48.3	65	111.2	184.5
2	25.7	50.9	70	121.4	200.4
4	27.4	53.6	75	132.2	217.1
6	29.2	56.3	80	143.6	234.9
8	31	59.2	85	155.7	253.7
10	32.8	62.2	90	168.4	273.5
12	34.8	65.2	95	181.8	294.4
14	36.8	68.3	100	195.9	316.4
16	38.8	71.5	105	210.8	339.6
18	40.9	74.9	110	226.4	364.1
20	43.1	78.3	115	242.8	389.9
22	45.3	81.8	120	260	416.9
24	47.6	85.4	125	278	445.4
			130	296.9	475.4

ABOUT THE AUTHOR

Norm Christopherson has worked in the HVAC industry since 1970. He is professor emeritus of HVAC at San Jose City College in California. He has custom designed and delivered courses for IBM, Coca Cola, Ford Motor Company, Advanced Micro Devices, Lockheed Martin, Loral Aerospace as well as for government agencies and numerous HVAC contractors. A national and local speaker for the Refrigeration Service Engineers Society, he has written several books and published dozens of HVAC articles. Norm is a Senior Training Specialist for Johnson Controls. He receives e-mail at nchristo@juno.com. Due to the high volume of e-mail he receives he reads all his mail but cannot personally respond to all the requests received. Reader suggestions for updates to this book are encouraged.

There is no substitute for experience and experience is no substitute for study.

– Norm Christopherson